A Primer for Financial Engineering

A Primer for Financial Engineering

Financial Signal Processing and Electronic Trading

Ali N. Akansu
New Jersey Institute of Technology
Newark, NJ

and

Mustafa U. Torun
Amazon Web Services, Inc.
Seattle, WA

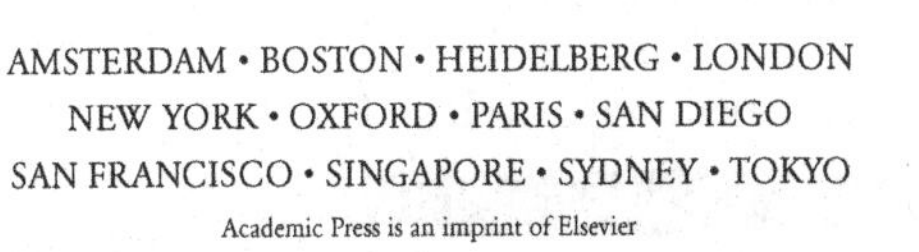
AMSTERDAM • BOSTON • HEIDELBERG • LONDON
NEW YORK • OXFORD • PARIS • SAN DIEGO
SAN FRANCISCO • SINGAPORE • SYDNEY • TOKYO
Academic Press is an imprint of Elsevier

Academic Press is an imprint of Elsevier
125 London Wall, London, EC2Y 5AS, UK
525 B Street, Suite 1800, San Diego, CA 92101-4495, USA
225 Wyman Street, Waltham, MA 02451, USA
The Boulevard, Langford Lane, Kidlington, Oxford OX5 1GB, UK

Library of Congress Cataloging-in-Publication Data
A catalog record for this book is available from the Library of Congress

British Library Cataloguing in Publication Data
A catalogue record for this book is available from the British Library

For information on all Academic Press publications
visit our website at http://store.elsevier.com/

ISBN: 978-0-12-801561-2

DEDICATION

To Daria, Fred and Irma

To Tuba and my parents

CONTENTS

PREFACE

This book presents the authors' professional reflections on finance, including their exposure to and interpretations of important problems historically addressed by experts in quantitative finance, electronic trading, and risk engineering. The book is a compilation of basic concepts and frameworks in finance, written by engineers, for a target audience interested in pursuing a career in financial engineering and electronic trading. The main goal of the book is to share the authors' experiences as they have made a similar transition in their professional careers.

It is a well recognized phenomenon on the Street that many engineers and programmers working in the industry are lacking the very basic theoretical knowledge and the nomenclature of the financial sector. This book attempts to fill that void. The material covered in the book may help some of them to better appreciate the mathematical fundamentals of financial tools, systems, and services they implement and are utilized by their fellow investment bankers, portfolio managers, risk officers, and electronic traders of all varieties including high frequency traders.

This book along with [1] may serve as textbook for a graduate level introductory course in Financial Engineering. The examples given in the book, and their MATLAB codes, provide readers with problems and project topics for further study. To access the MATLAB codes please visit the companion website http://booksite.elsevier.com/9780128015612/

The authors have benefited over the years from their affiliation with Prof. Marco Avellaneda of Courant Institute of Mathematical Sciences at the New York University. Thank you, Marco.

Ali N. Akansu
Mustafa U. Torun
February 2015

CHAPTER 1

Introduction

Financial engineers bring their knowledge base and perspectives to serve the financial industry for applications including the development of high-speed hardware and software infrastructure in order to trade securities (financial assets) within microseconds or faster, the design and implementation of high-frequency trading algorithms and systems, and advanced trading and risk management solutions for large size investment portfolios. A well-equipped financial engineer understands how the markets work, seeks to explain the behavior of the markets, develops mathematical and stochastic models for various signals related to the financial assets (such as price, return, volatility, comovement) through analyzing available financial data as well as understanding the market microstructure (studies on modeling the limit order book activity), then builds trading and risk management strategies using those models, and develops execution strategies to get in and out of investment positions in an asset. The list of typical questions financial engineers strive to answer include

- "What is the arrival rate of market orders and its variation in the limit order book of a security?"
- "How can one partition a very large order into smaller orders such that it won't be subject to significant market impact?"
- "How does the cross correlation of two financial instruments vary in time?"
- "Do high frequency traders have positive or negative impact on the markets and why?"
- "Can Flash Crash of May 6, 2010 happen again in the future? What was the reason behind it? How can we prevent similar incidents in the future?"

and many others. We emphasize that these and similar questions and problems have been historically addressed in overlapping fields such as finance, economics, econometrics, and mathematical finance (also known as quantitative finance). They all pursue a similar path of applied study. Mostly, the theoretical frameworks and tools of applied mathematics,

A Primer for Financial Engineering. http://dx.doi.org/10.1016/B978-0-12-801561-2.00001-0

statistics, signal processing, computer engineering, high-performance computing, information analytics, and computer communication networks are utilized to better understand and to address such important problems that frequently arise in finance. We note that financial engineers are sometimes called "quants" (experts in mathematical finance) since they practice quantitative finance with the heavy use of the state-of-the-art computing devices and systems for high-speed data processing and intelligent decision making in real-time.

Although the domain specifics of application is unique as expected, the interest and focus of a financial engineer is indeed quite similar to what a signal processing engineer does in professional life. Regardless of the application focus, the goal is to extract meaningful information out of observed and harvested signals (functions or vectors that convey information) with built-in noise otherwise seem random, to develop stochastic models that mathematically describe those signals, to utilize those models to estimate and predict certain information to make intelligent and actionable decisions to exploit price inefficiencies in the markets. Although there has been an increasing activity in the signal processing and engineering community for *finance applications* over the last few years (for example, see special issues of IEEE Signal Processing Magazine [2] and IEEE Journal of Selected Topics in Signal Processing [3], IEEE ICASSP and EURASIP EUSIPCO conference special sessions and tutorials on Financial Signal Processing and Electronic Trading, and the edited book *Financial Signal Processing and Machine Learning* [1]), inter-disciplinary academic research activity, industry-university collaborations, and the cross-fertilization are currently at their infancy. This is a typical phase in the inter-disciplinary knowledge generation process since the disciplines of interest go through their own learning processes themselves to understand and assess the common problem area from their perspectives and propose possible improvements. For example, speech, image, video, EEG, EKG, and price of a stock are all described as signals, but the information represented and conveyed by each signal is very different than the others by its very nature. In the foreword of Andrew Pole's book on statistical arbitrage [4], Gregory van Kipnis states "A description with any meaningful detail at all quickly points to a series of experiments from which an alert listener can try to reverse-engineer the [trading] strategy. That is why quant practitioners talk in generalities that are only understandable by the mathematically trained." Since one of the main goals of financial engineers is to profit from their findings of market inefficiencies complemented with expertise in trading, "talking in generalities" is understandable.

However, we believe, as it is the case for every discipline, financial engineering has its own "dictionary" of terms coupled with a crowded toolbox, and anyone well equipped with necessary analytical and computational skill set can learn and practice them. We concur that a solid mathematical training and knowledge base is a *must* requirement to pursue financial engineering in the professional level. However, once a competent signal processing engineer armed with the *theory of signals and transforms* and *computational skill set* understands the terminology and the finance problems of interest, it then becomes quite natural to contribute to the field as expected. The main challenge has been to understand, translate, and describe finance problems from an engineering perspective. The book mainly attempts to fill that void by presenting, explaining, and discussing the fundamentals, the concepts and terms, and the problems of high interest in financial engineering rather than their mathematical treatment in detail. It should be considered as an entry point and guide, written by engineers, for engineers to explore and possibly move to the financial sector as the specialty area. The book provides mathematical principles with cited references and avoids rigor for the purpose. We provide simple examples and their MATLAB codes to fix the ideas for elaboration and further studies. We assume that the reader does not have any finance background and is familiar with signals and transforms, linear algebra, probability theory, and stochastic processes.

We start with a discussion on market structures in Chapter 2. We highlight the entities of the financial markets including exchanges, electronic communication networks (ECNs), brokers, traders, government agencies, and many others. We further elaborate their roles and interactions in the global financial ecosystem. Then, we delve into six most commonly traded financial instruments. Namely, they are stocks, options, futures contracts, exchange traded funds (ETFs), currency pairs (FX), and fixed income securities. Each one of these instruments has its unique financial structure and properties, and serves a different purpose. One needs to understand the purpose, financial structure, and properties of such a financial instrument in order to study and model its behavior in time, intelligently price it, and develop trading and risk management strategies to profit from its *usually short lived* inefficiencies in the market. In Chapter 2, we also provide the definitions of a wide range of financial terms including buy-side and sell-side firms, fundamental, technical, and quantitative finance and trading, traders, investors, and brokers, European and American options, initial public offering (IPO), and others.

We cover the fundamentals of quantitative finance in Chapter 3. Each topic discussed in this chapter could easily be extended in an entire chapter of its own. However, our goal in Chapter 3 is to introduce the very basic concepts and structures as well as to lay the framework for the following chapters. We start with the price models and present continuous- and discrete-time geometric Brownian motion. Price models with local and stochastic volatilities, the definition of return and its statistical properties such as expected return and volatility are discussed in this chapter. After discussing the effect of sampling on volatility and price models with jumps, we delve into the modern portfolio theory (MPT) where we discuss the portfolio optimization, finding the best investment allocation vector for measured correlation (covariance/co-movement) structure of portfolio assets and targeted return along with its risk. Next, Section 3.4 revisits the capital asset pricing model (CAPM) that explains the expected return of a financial asset in terms of a risk-free asset and the expected return of the market portfolio. We cover various relevant concepts in Section 3.4 including the capital market line, market portfolio, and the security market line. Then, we revisit the relative value and factor models where the return of an asset is explained (regressed) by the returns of other assets or by a set of factors such as earnings, inflation, interest rate, and others. We end Chapter 3 by revisiting a specific type of factor that is referred to as eigenportfolio as detailed in Section 3.5.4. Our discussion on eigenportfolios lays the ground to present a popular trading strategy called statistical arbitrage (Section 4.6) in addition to filter the built-in market noise in the empirical correlation matrix of asset returns (Section 5.1.4).

As highlighted in Chapter 4, the practice of finance, traders, and trading strategies may be grouped in the three major categories. These groups are called fundamental, technical, and quantitative due to their characteristics. The first group deals with the financials of companies such as earnings, cash flow, and similar metrics. The second one is interested in the momentum, support, and trends in "price charts" of the markets. Financial engineers mostly practice quantitative finance, the third group, since they approach financial problems through mathematical and stochastic models, implementing and executing them by utilizing the required computational devices and trading infrastructure.

In contrast to investing into a financial asset (buying and holding a security for relatively long periods), trading seeks short-term price inefficiencies or trends in the markets. The goal in trading is simple. It is to buy low and sell high, and make profit coupled with a favorable risk level. Professional traders predefine and strictly follow a set of systematic rules (trading strategies) in analyzing the market data to detect investment opportunities as well as to intelligently decide how to react to those opportunities. In Chapter 4, we focus on quantitative (rules based) trading strategies. First, we present the terminology used in trading including long and short positions, buy, sell, short-sell, and buy-to-cover order types, and several others. We introduce the concepts like cost of trading, back-testing (a method to test a trading strategy using historical data), and performance measures for a trading strategy such as profit and loss (P&L) equitation and Sharpe ratio. Then, we cover the three most commonly used trading strategies. The first one is called pairs trading where the raw market data is analyzed to look for indicators identifying short lived relative price inefficiencies between a pair of assets (Section 4.5). The second one is called statistical arbitrage where the trader seeks arbitrage opportunities due to price inefficiencies across industries (Section 4.6). The last one is called trend following where one tracks strong upward or downward trends in order to profit from such a price move (Section 4.7). In the latter, we also discuss common trend detection algorithms and their ties to linear-time invariant filters. At the end of each section, we provide recipes that summarize the important steps of the given trading strategy. In addition, we also provide the MATLAB implementations of these strategies for the readers of further interest. We conclude the chapter with a discussion on trading in multiple frequencies where traders gain a fine grained control over the cycle of portfolio rebalancing process (Section 4.8).

Return and *risk* are the two *inseparable* and most important performance metrics of a financial investment. It is quite analogous with the two *inseparable* metrics of *rate* and *distortion* in rate-distortion theory [5]. In Section 3.3.1.2, we define the risk of a portfolio in terms of the correlation matrix of the return processes for the assets in the portfolio, $\mathbf{P}$. For a portfolio of N assets, there are $N(N-1)/2$ unknown cross-correlations and they need to be estimated through market measurements in order to form the empirical correlation matrix, $\hat{\mathbf{P}}$. It is a well-known fact that $\hat{\mathbf{P}}$ contains significant amount of inherent market noise. In Chapter 5, we revisit random matrix theory and leverage the asymptotically known behavior of the eigenvalues of random matrices in order to identify the noise component in $\hat{\mathbf{P}}$, and utilize eigenfiltering for its removal from measurements. Later in the chapter, we extend this method for the portfolios formed with statistical arbitrage (or any form of strategy that involves hedging of assets) (Section 5.1.4) and also for the case of trading in multiple frequencies (Section 5.2). The chapter includes various advanced methods for risk estimation (Section 5.3) like Toeplitz approximation to the empirical correlation matrix $\hat{\mathbf{P}}$ and use of discrete cosine transform (DCT) as an efficient replacement to Karhunen-Loéve transform (KLT) in portfolio risk management. Once the risk estimation is performed, the next step is to manage the risk. The main question to be addressed is "given that we know how to estimate the risk for the given investment allocation vector and correlation matrix of asset returns, how do we make a decision to change our positions in the assets in order to keep the investment risk at a predefined level?" We conclude the chapter with discussions on three different risk management methods. They are called stay in the ellipsoid (SIE), stay on the ellipsoid (SOE), and stay around the ellipsoid (SAE) as described in Section 5.4. (The locus of q_i, $1 \leq i \leq N$ satisfying (3.3.7) for a fixed value of risk σ_p is an ellipsoid centered at the origin. Hence, the names of these three methods include the word *ellipsoid*.)

Regardless of the trading strategy and risk management method utilized to come up with intelligent decisions on the investment allocation vector in time to rebalance the portfolio, the last and very important piece of the puzzle is to place orders and have them executed as originally planned. The market and limit order types are described in Chapter 6. The effect of orders on the current market price of an asset, called the market impact, and several algorithmic trading methods to mitigate it are discussed in this chapter. Then, we delve into market microstructure, examining the limit order book (LOB) and its evolution through placement of limit and market orders. After revisiting and commenting on important phenomenon in finance called Epps effect, the drop in the measured pairwise correlation between asset returns as the sampling (trading) frequency increases, we make some remarks on high frequency trading (HFT). We survey through publicly known HFT strategies and highlight the state-of-the-art on covariance estimation techniques with the high frequency data. We conclude the chapter with discussions on the low-latency (ultra high frequency) trading and impact of technology centric HFT on the financial markets.

The concluding remarks are presented in Chapter 7, followed by a reference list of books and articles cited in the book that may help the readers of interest for further study.

1.1 DISCLAIMER

The authors note that the material contained in this book is intended only for general information purpose and is not intended to be advice on any investment decision. The authors advise readers to seek professional advise before investing in financial products. The authors are not responsible or liable for any financial or other losses of any kind arising on the account of any action taken pursuant to the information provided in this book.

CHAPTER 2

Financial Markets and Instruments

We start with the definitions and descriptions of different entities in the financial markets and how they interact with each other. We introduce exchanges, electronic communication networks (ECNs), broker-dealers, market-makers, regulators, traders, and funds to better understand the financial ecosystem given that the legal framework, regulatory and compliance issues are beyond the scope of this book. In a separate section, we delve into the details of various types of financial instruments, i.e., stocks, options, futures, exchange traded funds (ETFs), currency pairs, and fixed income securities. Our goal in this chapter is not to provide the exhaustive details of these entities, instruments, and their relationships with each other. We rather aim to equip engineers with a good understanding of these concepts to navigate further in the area they choose to focus on through the references provided. We note that our primary focus is the financial markets and products offered in the United States. However, with only slight nuances, the concepts and definitions are globally applicable to any local financial market of interest.

2.1 STRUCTURE OF THE MARKETS

Financial instruments are bought and sold in venues called *exchanges*. There are many exchanges around the world. Most of them are specific to a particular class of financial instrument. New York Stock Exchange (NYSE), London Stock Exchange, and Tokyo Stock Exchange are some of the major *stock exchanges* around the world. Chicago Mercantile Exchange (CME), Chicago Board of Trade (CBOT), and London International Financial Futures and Options Exchange (LIFFE) are the largest *futures exchanges* in the world. Similarly, Chicago Board Options Exchange (CBOE) is the largest *options exchange* in the world. Exchanges are essentially *auction markets* in

A Primer for Financial Engineering. http://dx.doi.org/10.1016/B978-0-12-801561-2.00002-2

where parties *bid* and *ask* to *buy* and *sell* financial instruments, respectively. Traditionally, *traders* (people trading in their own accounts) and *brokers* (people trading in their clients' accounts) trade stocks and other instruments by being physically present in the exchange building, on the exchange floor. Hence, they are called *floor traders* and *floor brokers*. There are people on the exchange floor called *specialists* who are responsible for facilitating the trades, i.e., building the order book and matching the orders as well as maintaining *liquidity* of the product (*availability* of the financial product for buying and selling). When an investor wants to buy or sell a particular financial product, they call their banks or their brokers and *place the order*. Then, floor brokers are notified and they *execute the order* through the specialist, and *send back the confirmation*.

However, over time, the trading infrastructure and procedures have significantly changed with the introduction of ECNs. They are special computer networks for facilitating the execution of trades and carrying real-time market information to their consumers. ECNs provide access to real-time market data and let brokers and large traders trade with each other eliminating the need for a third party through *direct market access* (DMA). Some of the largest ECNs are Instinet, SelectNet by NASDAQ (National Association of Securities Dealers Automated Quotations), and NYSE Archipelago Exchange (NYSE ARCA).

The exchanges we have cited so far are very large in volume although they are just the tip of the iceberg in terms of the number of trading venues around the globe. Naturally, there is inter-exchange trading activity, flow of quotes for bids and asks for various types of products among those national and global venues. Just like exactly the same product may have different prices and sales volumes at two different grocery stores, the same financial product is usually traded at a different price, with different volume and liquidity, at different trading venues. This *fragmented market* structure makes it hard for small investors to trade at the best venue due to their lack of scale. *Broker-dealers* (BDs) facilitate trades for their clients in multiple venues through their sophisticated infrastructure comprised of the state-of-the-art low latency (high speed) data networking, high performance computing and storage facilities. Instead of having an account in each venue and build a costly trading infrastructure, the investor has one account with his broker that consolidates and offers most of these services for a fee. Brokers promise their clients to have their orders executed at the best price available in the market for a pre-defined transaction cost. Brokers compete among themselves by offering their clients low transaction costs, high liquidity, and access to many venues with lowest possible latency. Some brokers also provide sampled or near real-time market data to their clients for use in their *algorithmic trading* activity.

Investors, small or large scale, who randomly appear and disappear are not the only source of liquidity in the market. *Market makers* simultaneously place large buy and sell orders to maintain price robustness and stability in the *limit order book* (LOB) of a stock. Whenever there is an execution in either side, they immediately sell or buy from their own inventory or find an *offsetting order* to match. Market makers seek to profit from the price difference between the bid and ask orders they place that is called the *spread*. Since the size of the orders they place is usually much larger than the order size of a typical trader in the market, they in a sense, define, or *make the market* for that particular instrument.

Activity in the financial markets is regulated by government and non-governmental agencies. In the United States, the leading agency is the *Securities Exchange Commission* (SEC) which has the primary responsibility to enforce securities laws as well as to regulate the financial markets in the country. Moreover, there are agencies specific to a market or a small number of markets, such as the *Commodity Futures Trading Commission* (CFTC) that regulates the futures and options markets. In other countries, there are similar governmental and non-governmental bodies for the task. *International Organization for Securities Commission* (IOSCO) is responsible to regulate the securities and futures markets around the globe.

We have discussed, without any details, the basic concepts of facilitation and regulation of a trade. Although the main incentive to trade a financial instrument is to make profit by buying low and selling high, there are different types of players in the market based on their motivations to exploit the asset price inefficiencies and the way they trade. *Buy-and-hold* investors seek to profit in long term by holding the financial instrument for a long time, potentially years, and sell when the price is significantly high, leading to a high profit per trade over long time. On the other hand, *speculators* are not interested in holding the instrument for a long time and they seek to aggregate big profit from small profits on very large number trades. In the extreme case, there are *high frequency traders* in the market. Their holding times are as low as under a second or much less. Readers more interested are referred to Section 6.4 for a detailed discussion of *high-frequency trading* (HFT). There are also different types of processes and methods used to make a decision to trade a particular asset. *Fundamental traders*, *technical traders*, and *quantitative traders* (quants) employ different techniques to reach a decision to buy or sell an asset or a basket of assets. We present the foundations of these three trading categories in Chapter 4.

Only a small percentage of people manage their own money in the market through single or multiple investment accounts with broker-dealers. They are called *individual investors*. In contrast, most of the investors enter the market through investment *funds*. Funds are managed by finance professionals and serve as an investment vehicle for small and institutional investors. The fund managers make investment decisions on behalf of their clients through extensive financial analysis risk-return trade-offs combined with market intelligence and insights. Those investors who want to invest but do not know how to analyze the market and relevant risks of an investment commonly become a client of a *mutual fund* or a *hedge fund*. In general, mutual funds manage large amounts of investment capital, investing on behalf of many small and big investors. On the other hand, in general, hedge funds have smaller amount of capital under management invested by a lower number of investors. They usually invest in a wide variety of financial instruments. Traditionally speaking, hedge funds almost always employ hedged investment strategies to monitor and adjust the risk. However, today, a hedge fund often utilizes a large number of strategies comprised of hedged and non-hedged ones. Hedge funds are not open to investment for the general public. Therefore, they are less regulated compared to other funds.

It is quite common in the financial sector to refer to different market participants, e.g., investment banks and firms, analysts and institutions, as *buy side* and *sell side* entities. Basically, the difference between the two sides is about who is buying and who is selling the *financial services*. Sell side entities provide services such as financial advice, facilitation of trades, development of new products such as options (Section 2.2.2), futures contracts (Section 2.2.3), ETFs (Section 2.2.4), and many others. Such entities include investment banks, commercial banks, trade execution companies, and broker-dealers. In contrast, buy side entities buy those services in order to make profits from their investments on behalf of their clients. Some of these participants are hedge funds, mutual funds, asset management companies, and retail investors.

In the next section, we present the details of various types of most commonly traded financial instruments and highlight their specifics.

2.2 FINANCIAL INSTRUMENTS

We cover six types of the most popular financial instruments in this section. Namely, they are publicly traded company stocks, options, futures contracts, exchange traded funds (ETFs), currency pairs, and fixed income securities. We note that this list of instruments and our coverage of them are not exhaustive. However, this section will equip us with sufficient detail to understand different types of financial products and their distinctions.

2.2.1 Stocks

A *stock* represents a fraction of ownership in a company. It is delivered in the units of *shares*. However, it is also common to use the word "stock" rather than "stock share." Owner of a stock is called a *shareholder* of the company. A shareholder owns a percentage of the company determined by the ratio of the number of shares owned to the total outstanding shares. It is common that a company offers two types of stocks. Namely, they are *common stock* and *preferred stock*. A common stock entitles the owner a right to vote in the company whereas a preferred stock does not. However, a preferred stock has higher priority in receiving *dividends* and when the company files for bankruptcy. A dividend is a fraction of the profits delivered to the shareholders. Companies, as they make profit, may pay dividends to their shareholders throughout a year. Dividends can be paid in cash or in stocks.

Every stock is listed in a specific exchange such as NYSE or NASDAQ. However, they can be traded in multiple exchanges, ECNs, and possibly in dark pools as well. In order to identify stocks and standardize the naming method, every stock has a *symbol* or *ticker*. For example, the ticker for Apple Inc. stock is *AAPL* and the ticker for Google Inc. stock is *GOOG*. Moreover, there might be additional letters in the ticker of a stock, commonly separated by a dot, in order to differentiate different types of stocks (common, preferred, listed in US, listed in Mexico, etc.) a company offers to investors. For example, Berkshire Hathaway Inc. offers two different common stocks with the tickers BRK.A and BRK.B with their specific privileges.

Companies issue their stocks for the first time through an *initial public offering* (IPO) process. An IPO is assisted by an *underwriter*, a financial institution facilitating the IPO process. IPO is also referred to as *going public* since the company is not *privately held* anymore and its shares are publicly traded. A publicly traded company has to release a report on its financials and major business moves every quarter, and is subject to strict regulations. Therefore, there are still some very large companies that are privately held. Whether small or large, companies go public usually to raise cash, have liquidity in the market, offer stock options to their employees and attract talent, improve investor trust in them that may lead to reduced interest rates when they issue debts (Section 2.2.6), etc. Once publicly available, stocks can be purchased directly from the company itself, directly from an exchange via direct market access, or through a broker-dealer (most common).

In the old days, being listed in a major stock exchange was very important and prestigious for a company. However, it is changing as higher number of startups are going public. Stocks of those highly reputable and reliable companies with large *market capitalization* (market cap), the total value of the outstanding shares in the market, are sometimes called *blue chip* stocks (a reference to poker game). On the other extreme, stocks that do not meet the minimum criteria to be listed in stock exchanges are traded in the *over-the-counter* (OTC) markets. These stocks are called as *pink-sheets* (a reference to the color of the certificate) or *penny stocks*.

An *index* is the sum, or weighted sum, of stock prices for a *basket* (group) of stocks. They are mathematical formulas defining certain benchmarks to track market performance and they cannot be traded. The list of most widely quoted indices include Standard & Poor 500 (S&P 500), Dow Jones Industrial Average (DJIA), NASDAQ Composite (NASDAQ 100), Russell 1000, FTSE Eurotop 100, DAX, and Nikkei 225. There are many other indices with their focuses.

When the price of a stock gets very high, it is harder for small investors to buy and sell the stock, leading to reduced liquidity and possibly larger bid-ask spread (Section 6.2). In order to attract more liquidity, a company issues a *stock split*. For example, a 3-for-1 split delivers three *new shares* for each *old share* and simultaneously reducing the stock price by a factor of 3. Similarly, if the price of a stock goes very low, a company may issue a *reverse split*. As an example, a 1-for-4 reverse split delivers 1 *new share* for four *old shares* that increases stock price by a factor of 4. We note that splits do not change the market capitalization of a stock. Splits and dividends reveal themselves as impulses (jumps) in stock price and stock volume time series. Therefore, instead of using the raw historical price and volume data of a stock, it is preferred to use the adjusted price and volume that accounts for the splits and dividends. In Figure 2.2.1. we see closing price and adjusted price for Apple Inc. (APPL) stock.

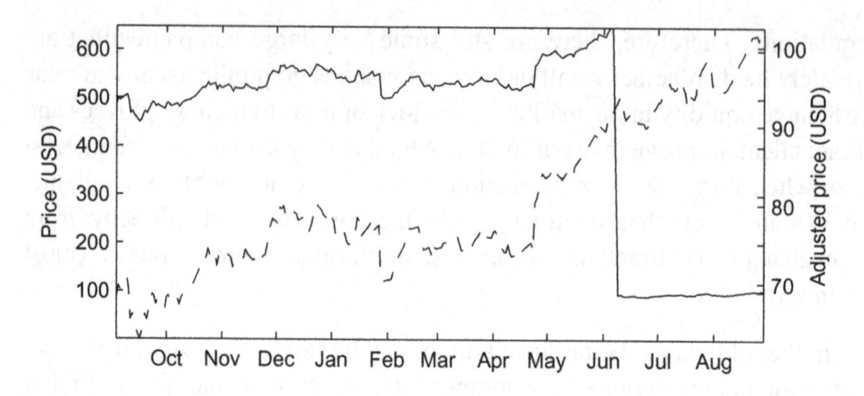

Figure 2.2.1 Daily closing price (solid line) and adjusted price (dashed line) for Apple Inc. (AAPL) stock between September 1, 2013 and August 31, 2014. Apple Inc. issued a 7-for-1 stock split on June 9, 2014, hence the large artificial drop in the closing price.

2.2.2 Options

An *option* is the right (not an obligation) to buy or sell a financial instrument at predefined price and terms. Underlying instruments may be stocks, bonds, market indices, commodities, bonds, and others. In this section, we focus on *stock options* for brevity and simplicity. In general, option contracts may be quite sophisticated and complex based on their design objectives. The buyer of a *call option (call)* has the right to buy the underlying stocks at the *strike price* and the seller of the call is obligated to deliver stocks if the buyer decides to *exercise the option* on or before the expiration date. Similarly, the buyer of a *put option (put)* has the right to sell the underlying stock at the strike price and the seller of the *put* is obligated to buy the stock if the buyer decides to exercise this right on or prior to its expiration. Although the transaction can be facilitated by a third party, option is a contract between the two parties, the seller and the buyer. In a sense, the seller writes a contract (legal document) and creates a tradable financial instrument. Therefore, seller of an option is also called the *writer* of the option. Options have their *expiration* (maturity) dates. A *European option* can be exercised only on its expiration date whereas an *American option* may be exercised at any time. In general, an option structured in US markets represents right to buy or sell 100 shares of the underlying stock.

Naturally, an option is not for free and it has its initial cost. Let S, K, and C be the spot price, strike price, and initial cost of buying an option, respectively. A long position in a call option (long call) pays off, call is *in-the-money*, only when $S > K + C$. Moreover, long call is *at-the-money* when $S = K + C$, and *out-of-the-money* when $S < K + C$. Similarly, a long put is in-the-money when $S < K - C$, at-the-money when $S = K - C$, and out-of-the-money when $S > K - C$. The payoff curves for long call, long put, short call (a short position in a call option), and short put are displayed in Figure 2.2.2. The buyer and the seller of the option is said to be in a long and short position in the option, respectively. We observe from Figure 2.2.2 that long call, long put, and short put have limited downside whereas downside for short call is unlimited. Similarly, all positions but the long call has limited upside. Long call and short put are *bullish* positions whereas long put and short put are *bearish* positions.

Options are mainly used for *hedging* and *leverage* of an open position in order to adjust the total risk. An investor can hedge his long position in a stock, to reduce the risk of his long position, by buying a put option on the same stock. Therefore, at the expiration date of the option, if the market (spot) price of the stock is less than the strike price of the option, investor can exercise the option and sell the stocks at a higher price than the spot price, hence, reduce his loss. Moreover, an option of a stock is usually much cheaper than its spot price. Therefore, instead of buying the shares of a stock, an investor can buy a larger number of put options with the same capital. In this scenario, spot price of the stock still needs to go up for the investor to

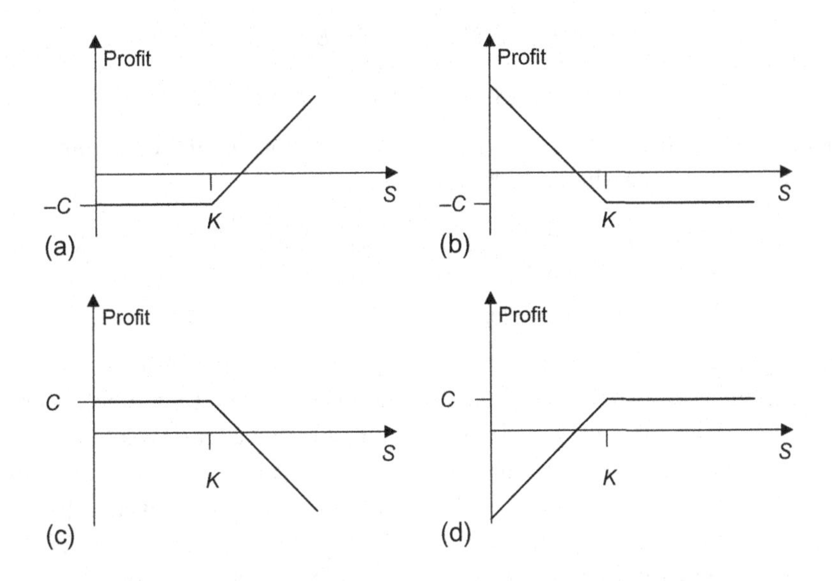

Figure 2.2.2 Payoff curves for (a) long call, (b) long put, (c) short call, and (d) short put. Long call, long put, and short put have limited downside whereas downside for short call is unlimited.

profit. However, the investor can generate much higher profit by exercising the large number of put options, rather than selling the stocks itself, hence, leveraging their capital.

Speculators in the market use combinations of the basic four positions in calls and puts and basic two positions in stocks to tailor investment strategies for a given risk target. The list of well-known strategies include the covered put, covered call, bull spread, bear spread, butterfly spread, iron condor, straddle, and strangle. For example, in a butterfly spread, an investor buys a call at strike price K_1, sells two calls at K_2, and sells a call at K_3 where $K_3 > K_2 > K_1$ all with the same expiration date. Butterfly spread allows the speculator to profit when the spot price is close to K_2 at the expiration date, and limits their loss otherwise.

Given their increased flexibility and complexity, options pricing is much more involved than stock pricing, and it has a very rich literature to learn from. In general, valuation models for options depend on the current spot price of the underlying asset, strike price, time to expiration date, and the volatility of the asset. Fischer Black and Myron Scholes developed their celebrated closed-form differential model for the price of a European option known as the *Black-Scholes options pricing formula* [6]. According to the formula, prices of a call and put option with spot price S, strike price K, and time-to-expiration date t are defined as

$$C(S,t) = F_{\mathrm{N}}(d_1)\,S - F_{\mathrm{N}}(d_2)\,Ke^{-rt}$$

$$P(S,t) = -F_{\mathrm{N}}(-d_1)\,S + F_{\mathrm{N}}(-d_2)\,Ke^{-rt}, \tag{2.2.1}$$

where $F_N(\cdot)$ is the cumulative distribution function (CDF) of a standard Gaussian random variable, r is the interest rate,

$$d_1 \triangleq \frac{1}{\sigma\sqrt{t}}\left[\ln\left(\frac{S}{K}\right) + \left(r + \frac{\sigma^2}{2}\right)t\right]$$

$$d_2 \triangleq d_1 - \sigma\sqrt{t},$$

and σ is the volatility of the underlying asset. We note that d_1 and d_2 are nothing else but defined parameters for ease of notation. The Black-Scholes model is often critiqued for impracticality since it assumes constant volatility and interest rate in time. Moreover, it is not easy in the model to account for the dividends paid for the underlying asset. However, it forms the basis for many other theoretical models including the *Heston model* [7] in which the volatility is not constant in time but a stochastic process. *Roll-Geske-Whaley* method is used to solve the Black-Scholes equation for an

American option with one dividend [8]. Binomial options pricing model developed by John Cox, Stephen Ross, and Mark Rubinstein known as the *Cox-Ross-Rubinstein* (CRR) model is used to model the price in the form of a binomial tree with each leaf corresponding to a possibility of the value at the expiration date. Granularity in CRR model can be adjusted to closely approximate the continuous-time Black-Scholes model. CRR is usually preferred over Black-Scholes model since it can model both European and American options and dividend payments with ease in the tree. Nevertheless, for most of the option classes, the differential equations in the models are intractable. In this case, one resorts to Monte Carlo simulation and finite difference methods to price the options.

2.2.3 Futures Contracts

A *forward contract* is an agreement between a seller and a buyer to deliver and purchase, respectively, a particular amount of an asset, *an underlying,* at a predefined price (delivery price) at a predefined date, *delivery date.* Traditionally, the underlying is a commodity such as wheat, oil, steel, silver, coffee beans, cattle, and many other tradable assets. However, underlying assets also include financial instruments such as currencies, treasury bonds, interest rates, stocks, market indices, and others. There are futures even for weather conditions, freight routes, hurricanes, carbon emissions, and movies.

Historically, forward contracts were used for balancing the supply and demand of the agricultural products such as wheat and cotton. Farmers would optimize their crops based on the contracts they have and the buyers would plan accordingly as both parties had a mutual contract to deliver/receive a product at a certain price at a certain date, hence, reducing their risks. Over time, this market had grown dramatically turning into a global *futures* market. The difference between a forward contract and a *futures contract* is that the latter is standardized, regulated, mostly traded in the exchanges, and cleared by financial institutions. Two large exchanges for futures are the Chicago Board of Trade (CBOT) and the London International Financial Futures and Options Exchange (LIFFE). In the Unites States, the futures market is regulated by the Commodity Futures Trading Commission (CFTC) and the National Futures Association (NFA).

The futures market is very liquid that makes it an attractive place to hedge investments and speculate their prices. There are still participants in the futures market that seek to buy and sell contracts to reduce their risk, *hedge*, for a particular commodity. However, similar to the options case, in today's global futures market, most of the trading activity is generated from the *speculation*. A great majority, almost all, of the futures contracts end without physical delivery of their underlying assets. Common speculation strategies are similar to those for options; going long, going short, or buying a spread of contracts. Typical spreads are calendar, inter-underlying, and inter-market in which the speculator buys and sells contracts for the same underlying with different delivery dates, for different underlying assets with the same expiration date, and for the same underlying and same expiration date in different exchanges, respectively.

In the futures markets, positions are settled every day. Therefore, in order to trade in the futures market, one needs a margin, cash in a bank account, for daily credits and debits due to potential gains and losses due to market fluctuations. However, margin requirements are relatively low, hence the *leverage* is high. Therefore, for the same amount of capital, one can bet for a larger number of underlying assets by buying the futures contract of the asset, rather than buying the asset itself.

Pricing of the futures contracts is fundamentally done by the assumption of no arbitrage. Let the delivery price, *forward price*, of a contract to be $F(T)$ and current spot price to be $S(t)$ where T is the time to delivery date. Then, the payoff for the seller of a contract is $F(T) - S(T)$ at the delivery. However, seller can *cash-and-carry* the underlying asset. Namely, he can own one unit of the asset and have a debt of $S(t)$ dollars, with the payoff amount $S(T) - S(t)\mathrm{e}^{r(T-t)}$ at the delivery where r is the interest rate. Hence, the total payoff of the seller is expressed as

$$F(T) - S(T) + S(T) - S(t)\mathrm{e}^{r(T-t)} = F(T) - S(t)\mathrm{e}^{r(T-t)}.$$

Due to the no arbitrage assumption, we have zero payoff. Therefore, the forward price of a contract is equal to

$$F(T) = S(t)\mathrm{e}^{r(T-t)}.$$

2.2.4 Exchange Traded Funds (ETFs)

ETF is a financial product that tracks a basket of other financial instruments or an index. They are similar to a mutual fund, however, they are traded just like a stock in the exchanges. Therefore, it is possible to invest in a basket or an index fund with very little capital or short-sell the fund to bet against an index. ETFs allow investors to diversify their portfolios with little cost. Namely, the only cost involved is the transaction cost paid to the broker-dealers. For example, in order to invest in the tech sector, intuitively, we would need to buy at least three large tech stocks such as Apple Inc. (AAPL), Microsoft Corporation (MSFT), and Google Inc. (GOOG). Ideally, we would need to buy all the stocks listed in the NASDAQ-100 index, a weighted average of the top 100 technology stocks based on market capitalization. Instead, we can simply buy PowerShares QQQ ETF which tracks the NASDAQ-100 index. Some ETFs are actively managed, their managers actively seek to beat the market. Holder of a managed ETF is charged a *management fee*. In general, an ETF has much smaller expense ratio than a mutual fund. The specifics of the basket and management fee for an ETF are found in its prospectus available by their issuers.

Net asset value (NAV) of a mutual fund is calculated at the end of each day whereas the price of an ETF changes in real time during the day as it is traded just like a stock in the exchanges. Therefore, they are also heavily used for speculation. However, buy and hold investors still gain performance close to the underlying index or basket as ETFs are rebalanced periodically to match performance over a long term. For example, at the time of writing, according to its prospectus, portfolio of QQQ is rebalanced quarterly and reconstituted annually.

Leveraged ETFs are higher risk instruments designed to match the *twice* or *triple performance* of an index. There are also *short leveraged ETFs* betting against indices. They target to match the *twice or triple opposite performance* of an index. The former and latter are useful for investors who believe the market is going to go up and down, respectively, and leverage their bets aggressively. However, it is common that some leveraged ETFs fail to deliver their predicted returns. See [9] for a detailed study on the path-dependence of the leveraged ETFs.

2.2.5 Currency Pairs

Foreign exchange markets, often abbreviated as *forex* or *FX*, are the largest in trading volume and capital in the world. Participants of these markets trade pairs of currencies by simultaneously buying a currency and selling the other. A simple example is the US Dollars and Japanese Yen pair (USD/JPY). In this example, USD is the base currency and the JPY is the counter currency. A quotation for USD/JPY as 109.34 means that 1 USD is equal to 109.34 JPY. In addition to the rate, the spread of a currency defined as the rate difference between buying and selling it is important. Spread is measured in *pips*. The pips is the smallest change for a rate. It usually is one percent of a percent. For example, a pip of 1 USD is 0.0001 USD. Brokers in FX market profit from the spread since there are no commissions. Moreover, the currency trader pays and receives interest on the currency sold and bought, respectively. The interest rate depends on the local country the traded currency belongs to. Therefore, the difference in interest rates for both legs of the currency pair is also an important factor in making FX trading decisions.

FX markets are decentralized and much less regulated compared to other markets due to their inherent cross-border market structure. However, currency brokers in the United States must be registered with the Futures Commission Merchants (FCMs) and they are regulated by CTFC. Currencies are traded worldwide. The majority of the trades occur in London and New York. The trading window is 24 hours during the business days. There is always high activity in some parts of the world during a day as the trading starts with Asia, moves on through Europe, North America, and back to Asia.

Participants in the FX markets use fundamental analysis, evaluating macro-economical and political factors as well as gross domestic product, manufacturing and sales, consumer price index of the countries and other economic factors to make trading decisions. Just like in stock markets or derivative markets, speculators also use technical analysis tools such as pivot points and Elliot waves for their trade decisions. Nevertheless, financial engineers and quants participating and performing research on FX markets apply their prior knowledge and experiences from the stock and derivative markets and develop more sophisticated arbitrage models tailored or tweaked for the currency pairs or a basket of currency pairs [10].

2.2.6 Fixed Income Securities

A *fixed income security* aims to deliver deterministic and mostly periodic returns. Bonds and money market securities are the most common types of fixed income securities. Investment in them brings very little risk (or no risk at all) but also delivers relatively low return compared to other financial instruments such as stocks. An entity like a government, a municipal authority, or a corporation issues bonds to lenders in order to borrow their capital with pre-defined terms and conditions. The issuer pays fixed interest to the lender, usually every 6 months, until the maturity date of the bond. Money market securities are considered as the cash market or short-term security market since they usually mature much faster than bonds.

Money market securities include but are not limited to treasury bills (T-Bills), certificates of deposit (CD), commercial papers, and repurchase agreements (repos). T-Bills are a way for governments to raise money from the public. Price for a T-Bill is less than its face (par) price. When it matures, government pays the full face price to the holder. Therefore, the deterministic return is the difference between the face value and the purchase value. Like bonds, T-Bills are generally purchased through a financial bidding process. They are considered as risk-free since they are backed by a government. However, the interest rate is often quite low. CDs are similar to T-Bills but they are offered by private banks. The interest rate is usually competitive but it is higher than that of a T-bill due to the increased risk of default. However, in the United States, CDs are insured, with certain limits, by Federal Deposit Insurance Corporation (FDIC) if the bank is a participant. Commercial papers are similar but they are offered by a corporation not a bank. By issuing commercial papers, companies raise cash quickly and avoid the banks.

In comparison to other markets, such as stock markets, the risk of money markets and bonds, hence, their return, is very little. Moreover, usually one needs larger capital to participate in the money market. Therefore, most of the players in the money market are actually subscribers of large mutual funds. In the stock market, brokers do not hold any risk due to the open positions created through them. The capital move is between the investor and the exchange. Brokerage business is merely to collect commissions as facilitators to access financial markets through exchanges. However, in the money market, there are no exchanges. Therefore dealers take the risk on their own account.

2.3 SUMMARY

There are many venues around the world where financial instruments are offered to investors and traded by market participants. Exchanges are auction markets in which different parties bid and ask to buy and sell instruments, and orders are matched by specialists. ECNs are special computer networks with state-of-the-art data processing power that allow broker-dealers and large traders to create a market ecosystem to interact and trade directly among themselves. In an ECN, the book keeping and order matching is done by machines rather than humans, mostly in sub-milliseconds if not in microseconds. *Market makers* with certain responsibilities and privileges in the exchange place large orders simultaneously bidding and asking for the same instrument. They seek profiting from the spread between the bid and ask prices. Other players in the market benefit from market makers since they provide liquidity. Financial markets are regulated mostly by government agencies. There are different types of investors based on their asset holding times. Namely, they are buy-and-hold investors, speculators, and high-frequency traders. Another category for traders is based on the methods they utilize in order to make their trade decisions. These trader types are called fundamental, technical, and quantitative traders. Sell side entities provide services to the buy side entities that make investment decisions on behalf of their clients in order to make profit by over-performing the market.

A stock represents a fraction of ownership in a company. An option is the right (not an obligation) to buy or sell a financial instrument at predefined price and time. A futures contract is an agreement between a seller and a buyer to deliver and to purchase, respectively, a particular amount of an asset at a predefined price and date. An exchange traded fund (ETF) is a financial product that tracks a basket of other financial instruments or a market index providing investors a low-cost diversification. Currency pairs are the largest contributors of trading activity where participants are seeking profit by exploiting the spread between two foreign currencies. A fixed income security delivers deterministic and mostly periodic returns.

CHAPTER 3

Fundamentals of Quantitative Finance

In this chapter, we revisit fundamental topics in quantitative finance including continuous- and discrete-time stock price models; stock price models with jumps; return, expected return, volatility, Sharpe ratio, and cross-correlation of assets; portfolio optimization, modern portfolio theory, and capital asset pricing model; relative value models, factor models, and eigen-portfolios. We discuss these concepts from a signal processing perspective along with several others. This chapter not only helps us to understand the fundamentals but also prepares us for the discussions presented in the following chapters.

3.1 STOCK PRICE MODELS

We briefly discussed the fundamental price models for options and futures contracts in Sections 2.2.2 and 2.2.3, respectively. In this section, we discuss in detail the fundamental models for the price of stocks.

3.1.1 Geometric Brownian Motion Model

Brownian motion, first discussed by Brown in 1827 in the context of motion of pollens, further explained by Einstein in 1905, and formulated by Wiener in 1918 has strong ties with the modeling of stock prices. Bachelier in 1900 described the price variation of a stock as a Brownian motion written as [11]

$$p(t) = p(0) + \mu t + \sigma w(t), \tag{3.1.1}$$

where $t \geq 0$ is the independent time variable, $p(t)$ is the price of the stock with an initial value $p(0)$, σ is the volatility, μ is the drift, and $w(t)$ is a Wiener process or standard Brownian motion such that $\mathrm{d}w(t)$ is a zero-mean

A Primer for Financial Engineering. http://dx.doi.org/10.1016/B978-0-12-801561-2.00003-4

and unit-variance Gaussian process, i.e., $dw(t) \sim \mathcal{N}(0,1)$. However, this model has problems. First, according to the model, it is possible for price to go below zero although stock prices are always positive. Moreover, according to the model, change in price over a time period is not a function of the initial price, $p(0)$. It suggests that stocks with different initial prices can have similar gains or losses in the same time interval. This is not the case in reality. For example, probability of observing a \$1 change in price over a day is less for a stock priced at \$10 than it is for a stock that is worth \$1,000. A better model for the stock price is the geometric Brownian motion in which the *rate of return* for a stock is defined as

$$\frac{dp(t)}{p(t)} = \mu dt + \sigma dw(t). \tag{3.1.2}$$

This stochastic differential equation has its analytic solution obtained by using Itô's Lemma [12] and expressed as

$$p(t) = p(0)\exp\left[\left(\mu - \frac{\sigma^2}{2}\right)t + \sigma w(t)\right]. \tag{3.1.3}$$

Before we discuss the proof, we provide three more definitions. The expected value and variance of $p(t)$ are expressed as, respectively,

$$E\{p(t)\} = p(0)e^{\mu t} \tag{3.1.4}$$

$$\operatorname{var}[p(t)] = p^2(0)e^{2\mu t}\left(e^{\sigma^2 t} - 1\right). \tag{3.1.5}$$

The log-price is defined as

$$\begin{aligned} s(t) &= \ln p(t) \\ &= \ln p(0) + \left(\mu - \frac{\sigma^2}{2}\right)t + \sigma w(t). \end{aligned} \tag{3.1.6}$$

Proof. We rewrite (3.1.2) as

$$dp(t) = \mu p(t)dt + \sigma p(t)dw(t). \tag{3.1.7}$$

Itô's Lemma states that if an Itô process $X(t)$ satisfies

$$dX(t) = \alpha[X(t),t]\,dt + \beta[X(t),t]\,dw(t), \tag{3.1.8}$$

where $\alpha[X(t),t]$ and $\beta[X(t),t]$ are two dimensional functions of $X(t)$ and t, then infinitesimal increment df for any function $f[X(t),t]$ differentiable in t and twice differentiable in $X(t)$ is given as

$$df = \left(\frac{\partial f}{\partial t} + \frac{\partial f}{\partial X(t)}\alpha + \frac{1}{2}\frac{\partial^2 f}{\partial X(t)^2}\beta^2\right)dt + \frac{\partial f}{\partial X(t)}\beta dw(t). \tag{3.1.9}$$

We do not display the independent variables $X(t)$ and t of functions $\alpha\,[X(t), t]$, $\beta\,[X(t), t]$, and $f[X(t), t]$ in (3.1.9) for ease of notation. Let

$$\begin{aligned} f[X(t), t] &= \ln X(t) \\ \alpha\,[X(t), t] &= \mu X(t) \\ \beta\,[X(t), t] &= \sigma X(t) \\ X(t) &= p(t). \end{aligned} \tag{3.1.10}$$

It follows from (3.1.9) and (3.1.10) that

$$\mathrm{d}\,[\ln p(t)] = \left(\mu - \frac{\sigma^2}{2}\right) \mathrm{d}t + \sigma \mathrm{d}w(t). \tag{3.1.11}$$

Since $\mathrm{d}w(t) \sim \mathcal{N}\,(0, 1)$, an infinitesimal increment in the log price (3.1.11) is a Gaussian with mean $\left(\mu - \sigma^2/2\right) \mathrm{d}t$ and variance $\sigma^2 \mathrm{d}t$. Since summation of the Gaussian random variables are also Gaussian it follows from (3.1.11) that $\ln p(t) - \ln p(0)$ is distributed as Gaussian with mean $\left(\mu - \sigma^2/2\right) t$ and variance $\sigma^2 t$. Therefore, we write

$$\ln p(t) - \ln p(0) = \left(\mu - \frac{\sigma^2}{2}\right) t + \sigma w(t). \tag{3.1.12}$$

Equivalently,

$$p(t) = p(0) \exp\left[\left(\mu - \frac{\sigma^2}{2}\right) t + \sigma w(t)\right], \tag{3.1.13}$$

which is identical to (3.1.3). Then, expected value of $p(t)$ is expressed as

$$E\,\{p(t)\} = p(0) E \left\{\exp\left[\left(\mu - \frac{\sigma^2}{2}\right) t + \sigma w(t)\right]\right\} \tag{3.1.14}$$

Recall that characteristic function of a Gaussian random variable is given as [13]

$$\Phi_X(\omega) = E\left\{\mathrm{e}^{j\omega X}\right\} = \int_{-\infty}^{\infty} f_X(x) \mathrm{e}^{j\omega x} dx = \mathrm{e}^{j\omega\eta - \nu^2\omega^2/2}, \tag{3.1.15}$$

where $f_X(x)$ is the probability density function of the Gaussian random variable $X \sim \mathcal{N}\left(\eta, \nu^2\right)$ and $j = \sqrt{-1}$ is the imaginary unit. Let $X(t) \triangleq \left(\mu - \sigma^2/2\right) t + \sigma w(t)$ and $j\omega \triangleq 1$. We know that $\left(\mu - \sigma^2/2\right) t + \sigma w(t)$ is distributed as Gaussian with mean $\left(\mu - \sigma^2/2\right) t$ and variance $\sigma^2 t$, i.e., $X \sim \mathcal{N}\left(\mu t - \sigma^2 t/2, \sigma^2 t\right)$. We have

$$\begin{aligned} E\left\{\mathrm{e}^X\right\} &= E\left\{\exp\left[\left(\mu - \frac{\sigma^2}{2}\right) t + \sigma w(t)\right]\right\} \\ &= \mathrm{e}^{\left(\mu t - \sigma^2 t/2\right) - \sigma^2 t(-1)/2} \\ &= \mathrm{e}^{\mu t}. \end{aligned} \tag{3.1.16}$$

From (3.1.14) and (3.1.16), we have

$$E\{p(t)\} = p(0)\mathrm{e}^{\mu t}, \tag{3.1.17}$$

which is identical to (3.1.4). Similarly we have

$$E\left\{p^2(t)\right\} = p^2(0)E\left\{\exp\left[2\left(\mu - \frac{\sigma^2}{2}\right)t + 2\sigma w(t)\right]\right\}. \tag{3.1.18}$$

For $j\omega \triangleq 2$ and $X \sim \mathcal{N}\left(\mu t - \sigma^2 t/2, \sigma^2 t\right)$, it follows from (3.1.15) that

$$\begin{aligned} E\left\{\mathrm{e}^{2X}\right\} &= E\left\{\exp\left[2\left(\mu - \frac{\sigma^2}{2}\right)t + 2\sigma w(t)\right]\right\} \\ &= \mathrm{e}^{2(\mu t - \sigma^2 t/2) - \sigma^2 t(-4)/2} \\ &= \mathrm{e}^{2\mu t + \sigma^2 t}. \end{aligned} \tag{3.1.19}$$

From (3.1.18) and (3.1.19), we have

$$E\left\{p^2(t)\right\} = p^2(0)\mathrm{e}^{2\mu t + \sigma^2 t}. \tag{3.1.20}$$

Finally, variance of $p(t)$ can be calculated from (3.1.17) and (3.1.20) as

$$\begin{aligned} \mathrm{var}\,[p(t)] &= E\left\{p^2(t)\right\} - E^2\{p(t)\} \\ &= p^2(0)\mathrm{e}^{2\mu t}\left(\mathrm{e}^{\sigma^2 t} - 1\right), \end{aligned}$$

which is identical to (3.1.5). □

3.1.2 Models with Local and Stochastic Volatilities

Geometric Brownian motion price model assumes a constant deviation of a stock return, i.e., constant volatility, σ. That assumption is not always realistic since the markets and prices of stocks are affected by various events that occur randomly and may last for a long time in some cases, e.g., an economical crisis, or appear and vanish within minutes, e.g., the *Flash Crash* of 2010 [14] (Section 6.4.4). Improved price models with local [15, 16] and stochastic [7, 17] volatilities take into account that the volatility itself is a function of time. In models with local volatility, the price model given in (3.1.2) is modified as

$$\frac{\mathrm{d}p(t)}{p(t)} = \mu \mathrm{d}t + \sigma\,[p(t), t]\,\mathrm{d}w(t), \tag{3.1.21}$$

where $\sigma\,[p(t), t]$ is the local volatility that depends on both time, t, and the price at time t, $p(t)$. On the other hand, models with stochastic volatility has the following form for the return

$$\frac{\mathrm{d}p(t)}{p(t)} = \mu \mathrm{d}t + \sigma(t)\,\mathrm{d}w_1(t), \tag{3.1.22}$$

where $\sigma(t)$, i.e., volatility as a function of time, is a random process and $\mathrm{d}w_1(t)$ is a normal process. One of the popular stochastic volatility models is the Heston [7] model in which volatility is a random process that satisfies the stochastic differential as given

$$d\sigma(t) = \kappa[\theta - \sigma(t)]\,\mathrm{d}t + \gamma\sqrt{\sigma(t)}\mathrm{d}w_2(t), \tag{3.1.23}$$

where κ is the mean-reversion speed, θ is the volatility in the long-term, γ is the volatility of the volatility $\sigma(t)$, and $\mathrm{d}w_2(t)$ is a normal process correlated with $\mathrm{d}w_1(t)$ given in (3.1.22). According to the Heston model, volatility is a mean-reverting process with a constant volatility and its infinitesimal changes are related to the ones of the price. We note that the model given in (3.1.23) is related to the celebrated Cox-Ingersoll-Ross process [18] used to model the short-term interest rates.

3.1.3 Discrete-Time Price Models and Return

It is a common practice to sample the price in time and to refer the stock returns with respect to their sampling periods, e.g., 30-min returns, 1-h returns, and end of day (EOD) returns. Discrete-time analog of geometric Brownian motion model is obtained by sampling the price with a certain time period as given

$$s(n) = s(n-1) + \mu + \sigma\xi(n), \tag{3.1.24}$$

where $s(n) = \ln p(n)$ is the log-price of a stock at discrete-time n with price $p(n)$, μ and σ are the *drift* and *volatility* of the stock, respectively, and $\xi(n)$ is the white Gaussian noise with $\xi(n) \sim \mathcal{N}(0,1)$. The log-return at discrete-time n is defined as

$$g(n) = s(n) - s(n-1) = \mu + \sigma\xi(n). \tag{3.1.25}$$

It follows from (3.1.24) and (3.1.25) that log-return is a Gaussian process with mean μ and variance σ^2, i.e., $g(n) \sim \mathcal{N}(\mu, \sigma^2)$. Moreover, it is a stationary and white noise process, i.e.,

$$E\{g(n-k)g(n-l)\} - \mu^2 = \sigma^2\delta_{k-l}. \tag{3.1.26}$$

3.2 ASSET RETURNS

Rate of return, or simply the *return* of an asset, is defined as the ratio of its price difference between the current and the previous samples, $p(n)$ and $p(n-1)$, over the previous price sample, $p(n-1)$, as given

$$r(n) = \frac{p(n) - p(n-1)}{p(n-1)} = \frac{p(n)}{p(n-1)} - 1. \tag{3.2.1}$$

For small values, return, $r(n)$, is an approximation to the log-return defined in (3.1.25), $g(n)$, due to the Taylor series expansion of the logarithm, i.e.,

$$g(n) = s(n) - s(n-1) = \ln\left[\frac{p(n)}{p(n-1)}\right] \cong \frac{p(n)}{p(n-1)} - 1 = r(n). \tag{3.2.2}$$

Since the value of return might get very small, it is customary in finance to use *basis points* (bps) instead of percent. One *bps* is one percent of a percent, i.e., 1 bps = 0.01%.

3.2.1 Expected Return, Volatility, and Cross-Correlation of Returns

Mean and standard deviation of the return of an asset, μ and σ, are referred to as *expected return* and *volatility* of an asset, defined as

$$\mu = E\{r(n)\}, \tag{3.2.3}$$

$$\sigma = \left(E\left\{r^2(n)\right\} - \mu^2\right)^{1/2}. \tag{3.2.4}$$

Expected return and volatility are the most basic measures for making investment decisions. If there were only two assets that investors can choose from, and only the expected returns and volatilities are known, a rational investor would choose to invest in the asset that has higher expected return as well as lower volatility. Lower volatility means lower investment risk. The *excess expected return* of an asset is the difference between expected return of the asset and the return of a risk-free asset, r_{f}, e.g., a treasury bill (Section 2.2.6). Ratio of the excess expected return to the volatility of an asset is called as the *Sharpe ratio*, defined as

$$\mathrm{SR} = \frac{\mu - r_{\mathrm{f}}}{\sigma}.$$

The higher the Sharpe ratio for an asset the higher the expected return for a given volatility.

Investors may further lower their risk via *diversification*, i.e., investing in a *portfolio* of assets. This practice requires us to know not only the mean and variance of the returns of individual assets, but also the cross-correlations of the asset returns in the portfolio. Cross-correlation (hence the covariance) of asset returns is an important aspect of modern portfolio theory [19] (Section 3.3), capital asset pricing model (Section 3.4), and relative value models (Section 3.5). Hence, it plays a central role in trading strategies such as pairs trading (Section 4.5) and statistical arbitrage (Section 4.6). The covariance of the returns of two assets, $r_1(n)$ and $r_2(n)$ with mean values $\mu_1(n)$ and $\mu_2(n)$, respectively, is defined as

$$\begin{aligned} \operatorname{cov}[r_1(n), r_2(n)] &= E\{r_1(n) - \mu_1\} E\{r_2(n) - \mu_2\} \\ &= E\{r_1(n) r_2(n)\} - \mu_1 \mu_2. \end{aligned} \tag{3.2.5}$$

Corresponding *cross-correlation coefficient* is defined as

$$\rho = \frac{\operatorname{cov}[r_1(n), r_2(n)]}{\sigma_1 \sigma_2} = \frac{E\{r_1(n) r_2(n)\} - \mu_1 \mu_2}{\sigma_1 \sigma_2}, \tag{3.2.6}$$

in the range $-1 \leq \rho \leq 1$. For the special cases where $\rho = 1$, $\rho = 0$, and, $\rho = -1$, asset returns are identical, have no correlation at all, and are completely opposite of each other, respectively.

Now, we extend the discussion into the scenario of multiple assets. We use matrices for the ease of notation and also to employ mathematical tools available in linear algebra. The return vector of size $N \times 1$ is defined as

$$\mathbf{r}(n) = \begin{bmatrix} r_1(n) & r_2(n) & \cdots & r_N(n) \end{bmatrix}^{\mathrm{T}}, \tag{3.2.7}$$

where $r_i(n)$ is the return of the ith asset, $1 \leq i \leq N$, and T is the matrix transpose operator. Expected return vector of size $N \times 1$ is defined as

$$\boldsymbol{\mu} = \begin{bmatrix} \mu_1 & \mu_2 & \cdots & \mu_N \end{bmatrix}^{\mathrm{T}}, \tag{3.2.8}$$

where μ_i is the expected return of the ith asset. Similarly, we define the diagonal volatility matrix of size $N \times N$ as

$$\boldsymbol{\Sigma} = \begin{bmatrix} \sigma_1 & . & . & 0 \\ 0 & \sigma_2 & . & 0 \\ & & & \\ . & . & \ddots & . \\ 0 & . & . & \sigma_N \end{bmatrix}, \tag{3.2.9}$$

where σ_i is the volatility of the ith asset. Since $\boldsymbol{\Sigma}$ is a diagonal matrix, we also have

$$\boldsymbol{\Sigma}^{-1} \boldsymbol{\Sigma} = \mathbf{I},$$

where $\mathbf{\Sigma}^{-1}$ is the inverse of matrix $\mathbf{\Sigma}$ and $\mathbf{I}$ is the $N \times N$ identity matrix defined as

$$\mathbf{I} = \begin{bmatrix} 1 & . & . & 0 \\ 0 & 1 & . & 0 \\ & & \ddots & \\ . & . & & . \\ 0 & . & . & 1 \end{bmatrix}. \tag{3.2.10}$$

We define the $N \times N$ covariance matrix of returns for an *N-asset portfolio* as

$$\begin{aligned} \mathbf{C} = [C_{ij}] &= \operatorname{cov}[r_i(n), r_j(n)] \\ &= E\{r_i(n)r_j(n)\} - \mu_i\mu_j. \end{aligned} \tag{3.2.11}$$

It is also possible to define $\mathbf{C}$ in terms of the return vector (3.2.7) and expected return vector (3.2.8) as follows

$$\mathbf{C} = E\left\{\mathbf{r}(n)\mathbf{r}^{\mathrm{T}}(n)\right\} - \boldsymbol{\mu}\boldsymbol{\mu}^{\mathrm{T}}.$$

Finally, we define the $N \times N$ correlation matrix as

$$\mathbf{P} = [P_{ij}] = \rho_{ij} = \frac{\operatorname{cov}[r_i(n), r_j(n)]}{\sigma_i\sigma_j} = \frac{E\{r_i(n)r_j(n)\} - \mu_i\mu_j}{\sigma_i\sigma_j}, \tag{3.2.12}$$

where $\rho_{ii} = 1$, i.e., all elements on the main diagonal of $\mathbf{P}$ are equal to one. Furthermore, $\mathbf{P}$ is a symmetric and positive definite matrix. From (3.2.9), (3.2.11), and (3.2.12), it follows that

$$\mathbf{C} = \mathbf{\Sigma}^{\mathrm{T}}\mathbf{P}\mathbf{\Sigma}.$$

Example 3.1. For a portfolio of two assets, we have the following *five-day* observations for their historical *end-of-day (EOD)* prices in US dollars

$$\begin{aligned} p_1(n) &= \{12.4, 12.8, 12.3, 12.1, 12.5\} \\ p_2(n) &= \{45.3, 45.9, 45.1, 44.7, 45.2\}, \end{aligned}$$

From (3.2.1), we have the *four-day* return processes for these two assets as

$$\begin{aligned} r_1(n) &= \{0.0323, -0.0391, -0.0163, 0.0331\} \\ r_2(n) &= \{0.0132, -0.0174, -0.0089, 0.0112\}. \end{aligned}$$

We estimate the (daily) expected asset returns (3.2.3) as $\hat{\mu}_1 = 24.98$ bps (24.98×10^{-4}) and $\hat{\mu}_2 = -4.67$ bps. Similarly, we estimate the (daily) asset volatilities (3.2.4) as $\hat{\sigma}_1 = 360.5$ bps and $\hat{\sigma}_2 = 150.8$ bps. The resulting daily Sharpe ratios are calculated as $\mathrm{SR}_1 = 0.069$ ($\mathrm{SR}_1 = 1.1$, annualized) and $\mathrm{SR}_2 = -0.031$ ($\mathrm{SR}_2 = -0.49$, annualized). We estimate covariance and cross-correlation coefficient as $\widehat{\operatorname{cov}}[r_1(n), r_2(n)] = 4 \times 10^{-4}$

and $\hat{\rho}_{12} = 0.7482$. Therefore, correlation and covariance matrices, $\hat{\mathbf{P}}$ and $\hat{\mathbf{C}}$, respectively, are found as

$$\hat{\mu} = \begin{bmatrix} 24.98 \times 10^{-4} & -4.67 \times 10^{-4} \end{bmatrix}^{\mathrm{T}},$$

$$\hat{\boldsymbol{\Sigma}} = \begin{bmatrix} 360.5 \times 10^{-4} & \\ & 150.8 \times 10^{-4} \end{bmatrix},$$

$$\hat{\mathbf{P}} = \begin{bmatrix} 1 & 0.7482 \\ 0.7482 & 1 \end{bmatrix},$$

$$\hat{\mathbf{C}} = \hat{\boldsymbol{\Sigma}}^{\mathrm{T}} \hat{\mathbf{P}} \hat{\boldsymbol{\Sigma}} = \begin{bmatrix} \hat{\sigma}_1^2 & \hat{\sigma}_1 \hat{\sigma}_2 \hat{\rho}_{12} \\ \hat{\sigma}_1 \hat{\sigma}_2 \hat{\rho}_{12} & \hat{\sigma}_2^2 \end{bmatrix} = \begin{bmatrix} 13 \times 10^{-4} & 4 \times 10^{-4} \\ 4 \times 10^{4} & 2 \times 10^{-4} \end{bmatrix}.$$

See file `returns.m` for the MATLAB code for this example.

3.2.2 Effect of Sampling Frequency on Volatility

It follows from (3.1.25) that we can write the log-price at discrete time n as a sum of initial log-price and all log-returns up to n as follows

$$s(n) = s(0) + \sum_{i=1}^{n} g(i). \tag{3.2.13}$$

If $s_{T_1}(n)$ and $s_{T_2}(n)$ are two discrete-time log-prices of the same asset, sampled with sampling periods $T_s = T_1$ and $T_s = T_2$, respectively, where $T_2 = kT_1$, $s_{T_2}(n) = s_{T_1}(kn)$, $k \in \mathbb{Z}$, and $k > 0$, then it follows from (3.1.25) and (3.2.13) that

$$\begin{aligned} g_{T_2}(n) &= s_{T_2}(n) - s_{T_2}(n-1) \\ &= s_{T_1}(kn) - s_{T_2}(kn - k) \\ &= s_{T_1}(0) + \sum_{i=1}^{kn} g_{T_1}(i) - s_{T_1}(0) - \sum_{i=1}^{kn-k} g_{T_1}(i) \\ &= \sum_{i=0}^{k-1} g_{T_1}(kn - i), \end{aligned} \tag{3.2.14}$$

where $g_{T_1}(n)$ and $g_{T_2}(n)$ are the log-returns associated with $s_{T_1}(n)$ and $s_{T_2}(n)$, respectively, via (3.2.13). Since the summation of Gaussian random variables is also a Gaussian random variable, if $g_{T_1}(n) \sim \mathcal{N}(\mu, \sigma^2)$ then $g_{T_2}(n) \sim \mathcal{N}(k\mu, k\sigma^2)$, and

$$\sigma_{T_2} = \sqrt{k}\sigma_{T_1}, \tag{3.2.15}$$

where σ_{T_1} and σ_{T_2} are the standard deviation of $g_{T_1}(n)$ and $g_{T_2}(n)$, respectively. Equality given in (3.2.15) shows that volatilities at different sampling frequencies of the same asset are related by square root of their sub-sampling ratio k.

3.2.3 Jumps in the Returns

Geometric Brownian motion model and its improved versions with local and stochastic volatilities discussed in Section 3.1 all have the continuity property in price. However, price of an asset is impacted by many reasons including asset specific and asset related business developments and financial news. Although some of those news are anticipated, there are many instances where these higher impact events happen quite randomly. We usually observe upward and downward abrupt price changes on any asset. These abrupt changes are referred to as *jumps* [20]. One of the simplest discrete-time price models with jumps is given as

$$s(n) = s(n-1) + j(n) + \xi(n), \tag{3.2.16}$$

where $j(n) \in \mathbb{R}$ is the abrupt price change, up or down, that happens at discrete-time n with certain statistical model, and $\xi(n) \sim \mathcal{N}(\mu, \sigma^2)$ is a Gaussian random noise process. We note that in (3.2.16), the random log-return $g(n)$ of (3.1.25) is modeled as the summation of two processes. Namely, a jump process $j(n)$, and a pure Gaussian noise process $\xi(n)$,

$$g(n) = j(n) + \xi(n) \tag{3.2.17}$$

In Figure 3.2.1a, realization of a Gaussian random process $\mathcal{N}(\mu, \sigma^2)$ with $\mu = 0.01$ bps and $\sigma = 2.11$ bps is shown. In Figure 3.2.1b, log-return of Apple Inc. (AAPL) stock on June 17, 2010 with a sampling period of $T_s = 5$ s is displayed. For this case, estimated mean (drift) and standard deviation

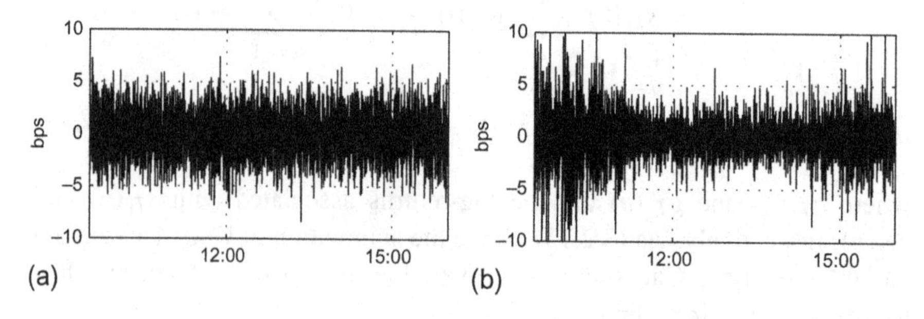

Figure 3.2.1 (a) A realization of a white Gaussian random process and (b) Log-returns of Apple Inc. (AAPL) stock on June 17, 2010 for sampling period $T_s = 5$ s.

(volatility) of the returns are 0.01 bps and 2.11 bps, respectively. We observe from Figure 3.2.1a and b that one needs to consider the jump process in the model in order to employ the basic price model more properly. Any jump of high significance is the main reason for the so-called regime change in the asset price.

In order to highlight the importance of jump processes in price modeling, we design a simple experiment as follows. From (3.2.15), we define the volatility estimation error between the two sampling intervals as follows

$$\varepsilon = \left| \hat{\sigma}(m)\sqrt{k/m} - \hat{\sigma}(k) \right|, \tag{3.2.18}$$

where $\hat{\sigma}(m)$ and $\hat{\sigma}(k)$ are the volatilities estimated at the intervals $T_s = m$ and $T_s = k$, respectively, via

$$\hat{\sigma}(T_s) = \left(\frac{1}{N-1} \sum_{i=0}^{N-1} \left[g_{T_s}(n-i) - \hat{\mu}(T_s) \right]^2 \right)^{1/2}, \tag{3.2.19}$$

$g_{T_s}(n)$ is the log-return of associated log-price sampled with the period T_s, $\hat{\mu}(T_s)$ is the estimated mean of the log-return as given

$$\hat{\mu}(T_s) = \frac{1}{N} \sum_{i=0}^{N-1} g_{T_s}(n-i), \tag{3.2.20}$$

and N is the estimation window length in samples. If the return process $g(n)$ in (3.2.17) were pure Gaussian, i.e., $g(n) = \xi(n)$, than the error term ε would be zero in accordance with (3.2.15). We employ a histogram based price jump detector where a return is labeled as a jump if its absolute value is larger than four times the estimated volatility, i.e., $4\hat{\sigma}$. Next, we define an artificial "jump-free" return process as

$$\hat{g}(n) = \hat{\xi}(n) = g(n) - \hat{j}(n). \tag{3.2.21}$$

Then, we calculate the volatility estimation error (3.2.18) for various sampling intervals spanning from $k = 1$ s to 300 s with $m = 1$ for both log-return and jump-free log-return of AAPL on day June 17, 2010, i.e., $g(n)$ and $\hat{g}(n)$ defined in (3.2.17) and (3.2.21), respectively. We calculate the error defined in (3.2.18) as a function of sampling interval k, $\varepsilon(k)$, and display it in Figure 3.2.2. We observe from the figure that removing jumps reduces the volatility estimation error. The jump process is an important phenomenon in the price formation, and one needs to take these abrupt changes into account for a better model. See, e.g., [20] for further details on jumps in asset returns.

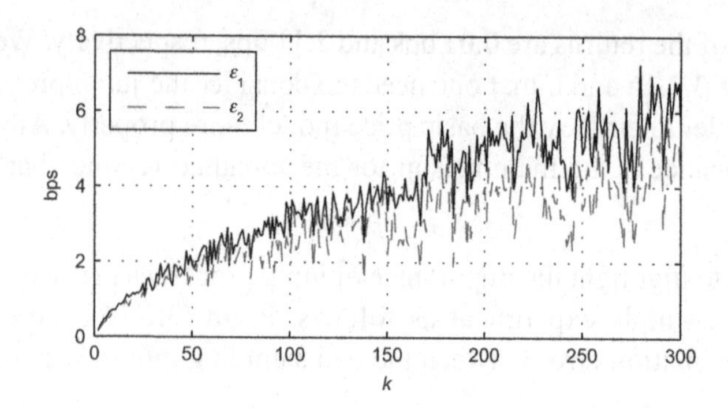

Figure 3.2.2 Volatility estimation error ε versus sampling period k with $m = 1$ as defined in (3.2.18) *for real and artificial (jump-free) returns of* (3.2.17) *and* (3.2.21), *i.e.,* ε_1 *and* ε_2*, respectively.*

3.3 MODERN PORTFOLIO THEORY

Modern portfolio theory (MPT) [19] provides a framework to create efficient portfolios with the minimized risk for a given expected return by optimally allocating the total investment capital among assets of the portfolio. Before providing the details of MPT, we first define the return and risk of portfolios.

3.3.1 Portfolio Return and Risk

Let us start with two-asset portfolio and extend the discussion for the case of multi-asset portfolio.

3.3.1.1 Two-Asset Portfolio

Portfolio return is the weighted average of the returns of the assets associated with it. Return of a two-asset portfolio is expressed as

$$r_{\mathrm{p}}(n) = q_1(n)r_1(n) + q_2(n)r_2(n), \tag{3.3.1}$$

where n is the discrete time variable, $q_i(n)$ is the ratio of the capital invested in the ith asset, and $r_i(n)$ is the return of the ith asset defined in (3.2.1). The investment amount, $q_i(n)$ in (3.3.1), can be dimensionless or its unit may be a currency. We omit the time index n in the following discussions for simplicity, noting that each term in an equation is a function of discrete time. Expected return of the two-asset portfolio is calculated as

$$\mu_{\mathrm{p}} = E\left\{r_{\mathrm{p}}\right\} = q_1 E\{r_1\} + q_2 E\{r_2\}. \tag{3.3.2}$$

Standard deviation of the portfolio return, i.e., the *risk* of the portfolio is given as

$$\sigma_{\mathrm{p}} = \left(E\left\{r_{\mathrm{p}}^2\right\} - E^2\left\{r_{\mathrm{p}}\right\}\right)^{1/2}$$
$$= \left(q_1^2\sigma_1^2 + 2q_1q_2\sigma_1\sigma_2\rho_{12} + q_2^2\sigma_2^2\right)^{1/2}, \qquad (3.3.3)$$

where σ_i is the volatility of ith asset and ρ_{ij} is the cross-correlation coefficient between the returns of ith and jth assets (3.2.6).

Example 3.2. For the assets given in Example 3.1, we calculate the return process of a portfolio with $q_1 = 0.7$ and $q_2 = 1 - q_1 = 0.3$ as

$$r_{\mathrm{p}}(n) = 0.7r_1(n) + 0.3r_2(n)$$
$$= \{0.0266, -0.0326, -0.0140, 0.0265\}.$$

Corresponding expected return and risk of the portfolio are calculated from (3.3.2) and (3.3.3) as $\hat{\mu}_{\mathrm{p}} = 16$ bps and $\hat{\sigma}_{\mathrm{p}} = 297$ bps, respectively. We observe that portfolio has lower expected return as well as volatility than the first and higher than the second asset. In Section 3.3.2, we investigate how one can do better than this random portfolio, by optimizing for the capital allocation coefficients q_1 and q_2. See file `two_asset_portfolio.m` for the MATLAB code of this example.

3.3.1.2 Multi-asset Portfolio

We extend the concepts of the previous section to the case of *N-asset* portfolio. Return of the *N-asset* portfolio is expressed as

$$r_{\mathrm{p}} = \mathbf{q}^{\mathrm{T}}\mathbf{r} = \sum_{i=1}^{N} q_i r_i, \qquad (3.3.4)$$

where $\mathbf{q}$ is the $N \times 1$ vector of capital allocation coefficients defined as

$$\mathbf{q} = \left[\begin{array}{cccc} q_1 & q_2 & \cdots & q_N \end{array}\right]^{\mathrm{T}}, \qquad (3.3.5)$$

and $\mathbf{r}$ is the return vector defined in (3.2.7). Hence, from (3.3.4), we calculate the expected return of the portfolio as

$$\mu_{\mathrm{p}} = E\left\{r_{\mathrm{p}}\right\} = \mathbf{q}^{\mathrm{T}}E\{\mathbf{r}\} = \mathbf{q}^{\mathrm{T}}\boldsymbol{\mu}, \qquad (3.3.6)$$

where elements of the $N \times 1$ vector $\boldsymbol{\mu}$ are the expected returns of individual assets defined in (3.2.8). Similarly, we obtain the risk of an N-asset portfolio as

$$\sigma_{\mathrm{p}} = \left(E\left\{r_{\mathrm{p}}^2\right\} - \mu_{\mathrm{p}}^2\right)^{1/2} \tag{3.3.7}$$

$$= \left(\mathbf{q}^{\mathrm{T}}\mathbf{C}\mathbf{q}\right)^{1/2} \tag{3.3.8}$$

$$= \left(\mathbf{q}^{\mathrm{T}}\boldsymbol{\Sigma}^{\mathrm{T}}\mathbf{P}\boldsymbol{\Sigma}\mathbf{q}\right)^{1/2} \tag{3.3.9}$$

$$= \left(\sum_{i=1}^{N}\sum_{j=1}^{N} q_i q_j \rho_{ij}\sigma_i\sigma_j\right)^{1/2} \tag{3.3.10}$$

where $\mathbf{C}$ is the covariance matrix of asset returns (3.2.11), $\boldsymbol{\Sigma}$ is the diagonal volatility matrix (3.2.9), and $\mathbf{P}$ is the correlation matrix (3.2.12).

3.3.2 Portfolio Optimization

In modern portfolio theory, portfolio optimization is achieved by minimizing the portfolio risk, σ_{p}, given in (3.3.7) with the constraint that the expected portfolio return, μ_{p}, of (3.3.6) is equal to a constant, i.e.,

$$\mu_{\mathrm{p}} = \mathbf{q}^{\mathrm{T}}\boldsymbol{\mu} = \sum_{i=1}^{N} q_i\mu_i = \mu. \tag{3.3.11}$$

There might be additional constraints such as constant investment capital of portfolio, i.e.,

$$\mathbf{q}^{\mathrm{T}}\mathbf{1} = \sum_{i=1}^{N} q_i = 1, \tag{3.3.12}$$

where $\mathbf{1}$ is an $N \times 1$ vector with all its elements equal to 1. The risk minimization problem to create an efficient portfolio subject to the constraints given in (3.3.11) and (3.3.12) can be solved by introducing two Lagrangian multipliers. We write the Lagrangian for this problem as

$$L\left(\mathbf{q}, \lambda_1, \lambda_2\right) = \frac{1}{2}\mathbf{q}^{\mathrm{T}}\mathbf{C}\mathbf{q} + \lambda_1\left(\mu - \mathbf{q}^{\mathrm{T}}\boldsymbol{\mu}\right) + \lambda_2\left(1 - \mathbf{q}^{\mathrm{T}}\mathbf{1}\right). \tag{3.3.13}$$

The optimum investment allocation vector is calculated by setting the partial derivatives of (3.3.13) to zero as follows

$$\frac{\partial L\left(\mathbf{q}, \lambda_1, \lambda_2\right)}{\partial \mathbf{q}} = 0, \quad \frac{\partial L\left(\mathbf{q}, \lambda_1, \lambda_2\right)}{\partial \lambda_1} = 0, \quad \frac{\partial L\left(\mathbf{q}, \lambda_1, \lambda_2\right)}{\partial \lambda_2} = 0.$$

These equations lead to the following solution for the optimum investment allocation vector

$$\mathbf{q}^* = \frac{\begin{vmatrix} \mu & \mathbf{1}^{\mathrm{T}}\mathbf{C}^{-1}\boldsymbol{\mu} \\ 1 & \mathbf{1}^{\mathrm{T}}\mathbf{C}^{-1}\mathbf{1} \end{vmatrix} \mathbf{C}^{-1}\boldsymbol{\mu} + \begin{vmatrix} \boldsymbol{\mu}^{\mathrm{T}}\mathbf{C}^{-1}\boldsymbol{\mu} & \mu \\ \boldsymbol{\mu}^{\mathrm{T}}\mathbf{C}^{-1}\mathbf{1} & 1 \end{vmatrix} \mathbf{C}^{-1}\mathbf{1}}{\begin{vmatrix} \boldsymbol{\mu}^{\mathrm{T}}\mathbf{C}^{-1}\boldsymbol{\mu} & \mathbf{1}^{\mathrm{T}}\mathbf{C}^{-1}\boldsymbol{\mu} \\ \boldsymbol{\mu}^{\mathrm{T}}\mathbf{C}^{-1}\mathbf{1} & \mathbf{1}^{\mathrm{T}}\mathbf{C}^{-1}\mathbf{1} \end{vmatrix}}, \tag{3.3.14}$$

We leave the derivation to the reader as an exercise. Set of optimum portfolios each satisfying the constraints of (3.3.11) and (3.3.12) for $-\infty < \mu < \infty$, form a curve in the (σ, μ) plane (Example 3.3). This curve is called the *Markowitz bullet.* Portfolios that lie on the upper-half of the Markowitz bullet are called efficient and they form the *efficient frontier*. Furthermore, only one of the efficient portfolios has the minimum risk, and therefore, it is called the *minimum risk portfolio*. The investment vector for the minimum risk portfolio is calculated by defining the Lagrangian as

$$L(\mathbf{q}, \lambda) = \frac{1}{2}\mathbf{q}^{\mathrm{T}}\mathbf{C}\mathbf{q} + \lambda\left(1 - \mathbf{q}^{\mathrm{T}}\mathbf{1}\right),$$

and setting its partial derivatives to zero as

$$\frac{\partial L(\mathbf{q}, \lambda)}{\partial \mathbf{q}} = 0, \quad \frac{\partial L(\mathbf{q}, \lambda)}{\partial \lambda} = 0.$$

These equations lead to the following solution for the investment allocation vector of the minimum risk portfolio

$$\mathbf{q}_{\min} = \frac{\mathbf{C}^{-1}\mathbf{1}}{\mathbf{1}^{\mathrm{T}}\mathbf{C}^{-1}\mathbf{1}}. \tag{3.3.15}$$

The minimum risk portfolio is unique. It has the minimum attainable risk, σ_{p}, coupled with the resulting expected return, μ_{p}.

Example 3.3. We have three assets in the portfolio with their correlation matrix (3.2.12) given as

$$\mathbf{P} = \begin{bmatrix} 1.0 & 0.6 & 0.2 \\ 0.6 & 1.0 & 0.7 \\ 0.2 & 0.7 & 1.0 \end{bmatrix}.$$

Expected returns of the assets are $\mu_1 = 0.07$, $\mu_2 = 0.03$, and $\mu_3 = 0.02$. Volatilities of the assets are $\sigma_1 = 0.02$, $\sigma_2 = 0.03$, and $\sigma_3 = 0.01$. We plot all the $(\sigma_{\mathrm{p}}, \mu_{\mathrm{p}})$ pairs calculated from (3.3.6) and (3.3.10) for the attainable portfolios, i.e., portfolios that satisfy the constraint of (3.3.12), in Figure 3.3.3. As expected, attainable portfolios are on or inside of the bullet. Moreover, we calculate the optimum investment allocation vectors

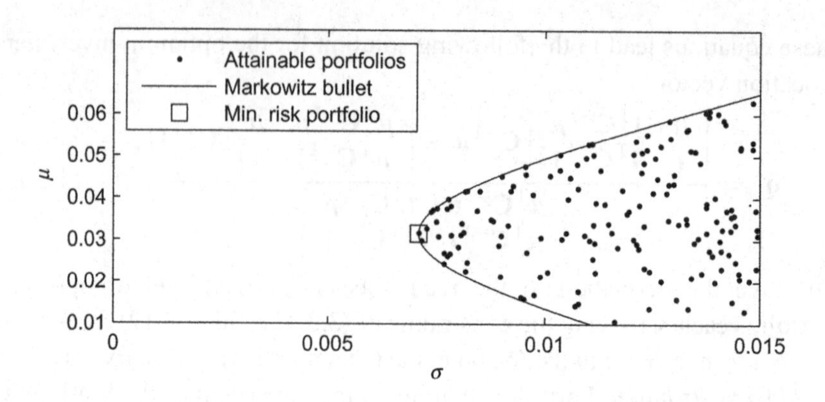

Figure 3.3.3 Markowitz bullet along with some of the attainable portfolios and the minimum risk portfolio. Portfolio consists of three assets. Their correlation structure, expected return, and volatilities are defined in Example 3.3.

by using (3.3.14) for different values of target average return of 3.3.11, and plot the Markowitz bullet in Figure 3.3.3. We note that the $(\sigma_{\min}, \mu_{\min})$ pair corresponding to $\mathbf{q}_{\min}$ in (3.3.15) is located at the far left tip of the Markowitz bullet as highlighted by a square in Figure 3.3.3. See file `markowitz.m` for the MATLAB code of this example.

3.4 CAPITAL ASSET PRICING MODEL

Capital asset pricing model (CAPM) explains the expected return of an asset, μ_i, in terms of the return of a risk-free (zero volatility) asset, r_f, such as interest on a treasury bill (Section 2.2.6) and expected return of the *market portfolio*, μ_M, as follows

$$\mu_i = r_f + \beta_i (\mu_M - r_f), \tag{3.4.1}$$

where β_i is referred to as the *beta* of the asset, i.e., a measure for the correlation of the asset return to the market return given as

$$\beta_i = \frac{\text{cov}(r_i, r_M)}{\sigma_M^2}. \tag{3.4.2}$$

where σ_M is the volatility of the market portfolio. CAPM was introduced by Sharpe [21]. Independently, Treynor [22], Lintner [23], and Mossin [24] also did similar work on the same subject. In this section, we start with adding a risk-free asset to the modern portfolio theory that leads us to the definitions of *capital market line* and *market portfolio*. Next, we derive the investment allocation vector of the market portfolio that leads us to obtain the beta given in (3.4.2). Finally, we discuss the volatility and expected return of an asset according to the CAPM.

3.4.1 Capital Market Line

Let us calculate the return of a portfolio, $\hat{p}$, in which we invest in a risk-free asset with return r_f (a constant) and a portfolio of risky assets, p,

$$r_{\hat{p}} = q_f r_f + q_p r_p,$$

where q_f and q_p are the capital allocation coefficients for the risk-free asset and portfolio p, respectively, where

$$q_f + q_p = 1, \tag{3.4.3}$$

and r_p is the return of portfolio p. Expected return of portfolio $\hat{p}$ is expressed as

$$E\{r_{\hat{p}}\} = \mu_{\hat{p}} = E\{q_f r_f + q_p r_p\} = q_f r_f + q_p \mu_p, \tag{3.4.4}$$

where μ_p is the expected return of portfolio p. The variance of the return for portfolio $\hat{p}$ is written as

$$\begin{aligned} \text{var}(r_{\hat{p}}) &= \text{var}(q_f r_f + q_p r_p) \\ &= q_f^2 \text{var}(r_f) + q_p^2 \text{var}(r_p) + 2 q_f q_p \text{cov}(r_f, r_p). \end{aligned}$$

Since r_f is a constant and its variance and covariance with any other return is zero, the last equation is reduced to

$$\text{var}(r_{\hat{p}}) = q_p^2 \text{var}(r_p).$$

Therefore, the risk of portfolio $\hat{p}$ is equal to

$$\sigma_{\hat{p}} = q_p \sigma_p. \tag{3.4.5}$$

From (3.4.3), (3.4.4), and (3.4.5) we deduce

$$\mu_{\hat{p}} - r_f = \frac{\mu_p - r_f}{\sigma_p} \sigma_{\hat{p}}, \tag{3.4.6}$$

which is the excess return over risk-free return achieved by investing in a risky portfolio along with the risk-free asset in $\hat{p}$. This excess return is also called the *risk premium*.

Equation in (3.4.6) represents a line in (σ, μ) plane. It follows from (3.4.6) that for a purely risk-free portfolio, i.e., $q_f = 1$ and $q_p = 0$, we have $(\sigma_{\hat{p}}, \mu_{\hat{p}}) = (0, r_f)$. Similarly, for a purely risky portfolio, i.e., $q_f = 0$

and $q_p = 1$, we have $(\sigma_{\hat{p}}, \mu_{\hat{p}}) = (\sigma_p, \mu_p)$. Moreover, the purely risk-free asset is located on the μ axis as it has zero risk, σ. On the other hand, the purely risky portfolio must be on or inside the Markowitz bullet. Any other portfolio with investment allocations in between the risk-free asset and the risky portfolio, i.e., different values of q_f and q_p, given that (3.4.3) holds, lies on the line defined in (3.4.6). One of the lines defined by (3.4.6) is called capital market line defined with the help of the following example.

Example 3.4. We continue Example 3.3 and we display two possible lines defined by (3.4.6) for different risky portfolios in Figure 3.4.4 for $r_f = 0.015$. We observe from the figure that all the portfolios lying on the line that is tangent to the Markowitz efficient frontier have higher expected return, μ, for a given risk, σ. Therefore, all rational investors would choose this line. This tangent line is called the *capital market line* (CML). The portfolio at the point of tangency (highlighted by a triangle) is called the *market portfolio.* See file `cml.m` for the MATLAB code for this example.

3.4.2 Market Portfolio

Let us derive the investment allocation vector of the market portfolio. Slope of the CML given in (3.4.6) is equal to

$$\frac{\mu_p - r_f}{\sigma_p}. \tag{3.4.7}$$

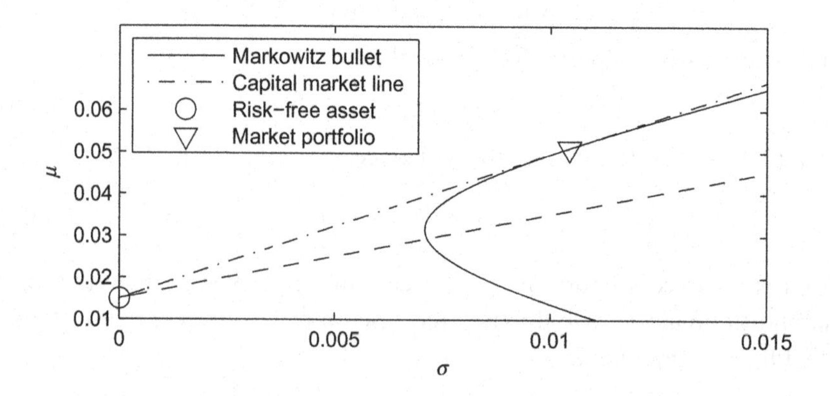

Figure 3.4.4 Markowitz bullet along with an arbitrary line for $\mu_{\hat{p}} = 0.036$ and $\sigma_{\hat{p}} = 0.0104$ in (3.4.6) and the capital market line (CML), i.e., $\mu_{\hat{p}} = \mu_M$ and $\sigma_{\hat{p}} = \sigma_M$. CML is tangent to the Markowitz efficient frontier and the tangency point is called the market portfolio. We assume the risk free asset has a return $r_f = 0.015$ as given in Example 3.4. Correlation structure, expected return, and volatilities of the three assets in the portfolio are defined in Example 3.3.

It follows from (3.3.8) and (3.3.11) that for any portfolio that is on or inside the Markowitz bullet, the slope is equal to

$$\frac{\mathbf{q}^{\mathrm{T}}\boldsymbol{\mu}-r_{\mathrm{f}}}{\left(\mathbf{q}^{\mathrm{T}}\mathbf{C}\mathbf{q}\right)^{1/2}}.$$

Market portfolio is the one that maximizes this slope subject to the constraint that $\mathbf{q}^{\mathrm{T}}\mathbf{1} = 1$. Similar to the portfolio optimization problem discussed earlier, we define the Lagrangian as follows

$$L(\mathbf{q},\lambda)=\frac{\mathbf{q}^{\mathrm{T}}\boldsymbol{\mu}-r_{\mathrm{f}}}{\left(\mathbf{q}^{\mathrm{T}}\mathbf{C}\mathbf{q}\right)^{1/2}}+\lambda\left(1-\mathbf{q}^{\mathrm{T}}\mathbf{1}\right),$$

and set its partial derivatives to zero, i.e.,

$$\frac{\partial L(\mathbf{q},\lambda)}{\partial \mathbf{q}}=0,\quad \frac{\partial L(\mathbf{q},\lambda)}{\partial \lambda}=0.$$

These equations lead us to the following solution for the investment allocation vector of the market portfolio

$$\mathbf{q}_{\mathrm{M}}=\frac{\mathbf{C}^{-1}\left(\boldsymbol{\mu}-r_{\mathrm{f}}\mathbf{1}\right)}{\mathbf{1}^{\mathrm{T}}\mathbf{C}^{-1}\left(\boldsymbol{\mu}-r_{\mathrm{f}}\mathbf{1}\right)}. \tag{3.4.8}$$

We leave the derivation to the reader as an exercise. Hence, return of the market portfolio can be calculated as

$$r_{\mathrm{M}}=\mathbf{q}_{\mathrm{M}}^{\mathrm{T}}\mathbf{r}, \tag{3.4.9}$$

where $\mathbf{r}$ is the $N \times 1$ vector of asset returns defined in (3.2.7). Similarly, the expected return, μ_{M}, and risk, σ_{M}, of the market portfolio are given as

$$\mu_{\mathrm{M}}=\mathbf{q}_{\mathrm{M}}^{\mathrm{T}}\boldsymbol{\mu}, \tag{3.4.10}$$

$$\sigma_{\mathrm{M}}=\left(\mathbf{q}_{\mathrm{M}}^{\mathrm{T}}\mathbf{C}\mathbf{q}_{\mathrm{M}}\right)^{1/2}. \tag{3.4.11}$$

Example 3.5. We continue from Example 3.4 and calculate the investment allocation vector for the market portfolio using (3.4.8) as

$$\mathbf{q}_{\mathrm{M}}=\left[\begin{array}{lll}0.7 & -0.39 & 0.69\end{array}\right]^{\mathrm{T}}, \tag{3.4.12}$$

in which $\mathbf{q}_{\mathrm{M}}^{\mathrm{T}}\mathbf{1} = 1$ as expected. Corresponding expected return and risk of the market portfolio are calculated from (3.4.10) and (3.4.11) as $\mu_{\mathrm{M}} = 0.051$ and $\sigma_{\mathrm{M}} = 0.01$, respectively. We note that the "market portfolio" in this example consists of three assets, the same as in Example 3.3. In general, we study MPT for a large portfolio that is comprised of many assets from various industries. Therefore, in that sense, the use of the word "market" makes more sense. See file market_portfolio.m for the MATLAB code of this example.

3.4.3 Beta of an Asset

We model the return process of an asset as a function of the return process of the market portfolio as follows

$$r_i = \alpha_i + \beta_i r_{\mathrm{M}} + \epsilon_i, \tag{3.4.13}$$

where α_i and β_i are constants, r_i and r_{M} are the return processes for the ith asset and the market portfolio given in (3.4.9), respectively, and ϵ_i is the prediction error process. Equation (3.4.13) is nothing else but a linear regression of the asset returns with respect to the market portfolio returns. The coefficient β_i is referred to as the *beta* of the asset. It is used as a measure of the correlation between returns of an asset and the market. Let us substitute (3.4.13) in (3.2.5) as follows

$$\mathrm{cov}\,(r_i, r_{\mathrm{M}}) = E\left\{r_i r_{\mathrm{M}}\right\} - \mu_i \mu_{\mathrm{M}}. \tag{3.4.14}$$

where μ_{M} is the expected return of the market portfolio as given in (3.4.10). Moreover, from (3.4.13) we obtain the last term as

$$\mu_i \mu_{\mathrm{M}} = \alpha_i \mu_{\mathrm{M}} + \beta \mu_{\mathrm{M}}^2 + E\left\{\epsilon_i\right\} \mu_{\mathrm{M}}.$$

Therefore, substitution of (3.4.13) into (3.4.14) yields

$$\begin{aligned} \mathrm{cov}\,(r_i, r_{\mathrm{M}}) &= E\left\{(\alpha_i + \beta_i r_{\mathrm{M}} + \epsilon_i)\, r_{\mathrm{M}}\right\} - \mu_i \mu_{\mathrm{M}} \\ &= \beta_i E\left\{r_{\mathrm{M}}^2\right\} - \beta_i \mu_{\mathrm{M}}^2. \end{aligned} \tag{3.4.15}$$

It follows from (3.4.15) that

$$\beta_i = \frac{\mathrm{cov}\,(r_i, r_{\mathrm{M}})}{\sigma_{\mathrm{M}}^2} = \rho_{i,M} \frac{\sigma_i}{\sigma_{\mathrm{M}}}, \tag{3.4.16}$$

where $\rho_{i,\mathrm{M}}$ and $\sigma_{\mathrm{M}} = (E\left\{r_{\mathrm{M}}^2\right\} - \mu_{\mathrm{M}}^2)^{1/2}$ are the cross-correlation coefficient of r_i and r_{M}, and risk of the market portfolio as expressed in (3.2.6) and (3.4.11), respectively. We discuss the effect of beta of an asset on its volatility and expected return next.

3.4.4 Volatility in CAPM

Let us express the volatility of the ith asset with the return given in (3.4.13) in terms of beta, β_i and error, ϵ_i. The variance of the return process is equal to

$$
\begin{aligned}
\operatorname{var}(r_i) &= \operatorname{var}(\alpha_i + \beta_i r_{\mathrm{M}} + \epsilon_i) \\
&= \operatorname{var}(\beta_i r_{\mathrm{M}} + \epsilon_i) \\
&= \beta_i^2 \operatorname{var}(r_{\mathrm{M}}) + \operatorname{var}(\epsilon_i) \\
&= \beta_i^2 \sigma_{\mathrm{M}}^2 + \sigma_{\epsilon_i}^2, \qquad (3.4.17)
\end{aligned}
$$

that follows from the assumptions that α_i is a constant, and covariance of the error to the returns of the market portfolio is zero, i.e., $\operatorname{cov}(r_{\mathrm{M}}, \epsilon) = 0$. Since the volatility is the standard deviation of return, from (3.4.17), we have the volatility of the ith asset as

$$
\sigma_i = \left(\beta_i^2 \sigma_{\mathrm{M}}^2 + \sigma_{\epsilon_i}^2\right)^{1/2}. \qquad (3.4.18)
$$

We observe from (3.4.18) that there are two contributors to the volatility of the ith asset. The first term, $\beta_i^2 \sigma_{\mathrm{M}}^2$, is a function of the market portfolio and the beta of the asset. Therefore, it is referred to as the *systematic risk* of the asset. The second term, σ_ϵ^2, has nothing to do with the market portfolio and it is only a function of the error in the regression given in (3.4.13). Therefore, it is referred to as the *unsystematic risk* or the *idiosyncratic* risk of the asset.

3.4.5 Expected Return in CAPM

According to the CAPM, expected return of the ith asset is equal to

$$
\mu_i = r_{\mathrm{f}} + \beta_i (\mu_{\mathrm{M}} - r_{\mathrm{f}}). \qquad (3.4.19)
$$

which states that expected return of an asset is related to the expected return of the market portfolio with its beta parameter.

Proof. In order to prove (3.4.19), let us form a portfolio by investing q of our capital to the market portfolio and $1 - q$ of our capital to the ith asset, $0 \leq q \leq 1$. The return of this particular portfolio is given as

$$
r_{\mathrm{p}} = q r_{\mathrm{M}} + (1 - q) r_i, \qquad (3.4.20)
$$

with mean and variance expressed as

$$
\begin{aligned}
\mu_{\mathrm{p}} &= q \mu_{\mathrm{M}} + (1 - q) \mu_i, \qquad (3.4.21) \\
\sigma_{\mathrm{p}}^2 &= q^2 \sigma_{\mathrm{M}}^2 + (1 - q)^2 \sigma_i^2 + 2q(1 - q) \operatorname{cov}(r_{\mathrm{M}}, r_i). \qquad (3.4.22)
\end{aligned}
$$

For each value of q, there is a point on the (σ, μ) plane. In one extreme case when $q = 0$, we have $(\sigma_{\mathrm{p}}, \mu_{\mathrm{p}}) = (\sigma_i, \mu_i)$. In the other extreme when $q = 1$, we have $(\sigma_{\mathrm{p}}, \mu_{\mathrm{p}}) = (\sigma_{\mathrm{M}}, \mu_{\mathrm{M}})$, that is the market portfolio. The set of points $(\sigma_{\mathrm{p}}, \mu_{\mathrm{p}})$ for all the values of $0 \leq q \leq 1$ form a smooth curve in the (σ, μ) plane as displayed in Figure 3.4.5 for an arbitrary asset. Moreover, $(\sigma_{\mathrm{M}}, \mu_{\mathrm{M}})$

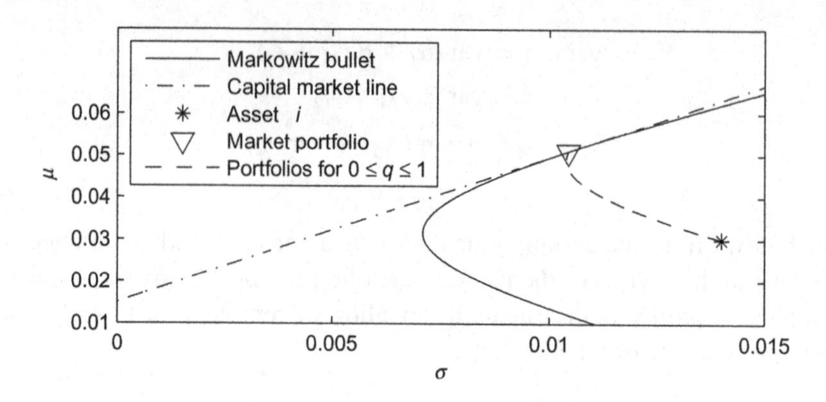

Figure 3.4.5 Markowitz bullet along with capital market line. Portfolios with the return defined in (3.4.20) form a curve on (σ, μ) plane. Special cases are at the boundaries, i.e., the asset i ($q = 0$) and market portfolio ($q = 1$). We assume the expected return, volatility, and beta of the asset with the market portfolio, are $\mu_i = 0.03$, $\sigma_i = 0.014$, and $\beta_i = 0.9$, respectively. Correlation structure, expected return, and volatilities of the three assets in the portfolio are defined in Example 3.3.

is the tangency point for this curve, efficient frontier, and CML. Therefore, the slope of this smooth curve is equal to the slope of CML given in (3.4.7) when $q = 1$, i.e.,

$$\left.\frac{\partial \mu_{\mathrm{p}}}{\partial \sigma_{\mathrm{p}}}\right|_{q=1} = \frac{\mu_{\mathrm{M}} - r_{\mathrm{f}}}{\sigma_{\mathrm{M}}}. \tag{3.4.23}$$

From chain rule in calculus, we have

$$\frac{\partial \mu_{\mathrm{p}}}{\partial \sigma_{\mathrm{p}}} = \frac{\partial \mu_{\mathrm{p}}/\partial q}{\partial \sigma_{\mathrm{p}}/\partial q}. \tag{3.4.24}$$

We calculate the derivative of μ_{p} given in (3.4.21) with respect to q as

$$\frac{\partial \mu_{\mathrm{p}}}{\partial q} = \mu_{\mathrm{M}} - \mu_i. \tag{3.4.25}$$

Similarly, we calculate the derivative of σ_{p} using σ_{p}^2 given in (3.4.22) with respect to q as

$$2\sigma_{\mathrm{p}}\frac{\partial \sigma_{\mathrm{p}}}{\partial q} = 2q\sigma_{\mathrm{M}}^2 - 2(1-q)\sigma_i^2 + 2(1-2q)\operatorname{cov}(r_{\mathrm{M}}, r_i). \tag{3.4.26}$$

From (3.4.23), (3.4.24), (3.4.25), and (3.4.26), we have

$$\left.\frac{2\sigma_{\mathrm{p}}(\mu_{\mathrm{M}} - \mu_i)}{2q\sigma_{\mathrm{M}}^2 - 2(1-q)\sigma_i^2 + 2(1-2q)\operatorname{cov}(r_{\mathrm{M}}, r_i)}\right|_{q=1} = \frac{\mu_{\mathrm{M}} - r_{\mathrm{f}}}{\sigma_{\mathrm{M}}}. \tag{3.4.27}$$

Given that $\sigma_{\text{p}} = \sigma_{\text{M}}$ for $q = 1$ and beta of the asset i, β_i, given in (3.4.16), (3.4.27) can be further simplified to

$$\frac{\mu_{\text{M}} - \mu_i}{1 - \beta_i} = \mu_{\text{M}} - r_{\text{f}},$$

that is identical to (3.4.19). □

3.4.6 Security Market Line

In (3.4.1), given that r_{f} and μ_{M} are known, the expected return of the ith asset is a linear function of its beta, $\mu_i(\beta_i)$, written as

$$\mu_i(\beta_i) = r_{\text{f}} + \beta_i (\mu_{\text{M}} - r_{\text{f}}). \tag{3.4.28}$$

This line in the (β, μ_i) plane is referred to as *security market line* (SML). As we stated before, the higher the correlation with the market portfolio, the higher the beta, and the higher the systematic risk. Therefore, SML displays the expected return of investing in asset i for a given systematic risk. In the special cases, for $\beta = 1$, we have $\mu_i(1) = \mu_{\text{M}}$ that means asset i has perfect correlation with the market and investing in it is identical to investing into the market portfolio. For $\beta = 0$, we have $\mu_i(0) = r_{\text{f}}$ that means asset i has no correlation with the market and investing in it is identical to investing in the risk-free asset. We display the SML for an arbitrary risk-free asset and market portfolio in Figure 3.4.6.

However, for some periods in time, the expected return of an asset can be above or below the SML in the (β, μ_i) plane dictated by the linear regression

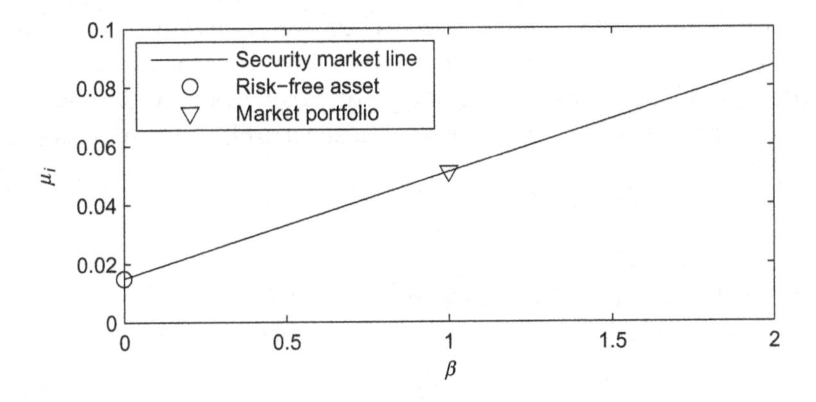

Figure 3.4.6 Security market line (SML) of the ith asset as defined in (3.4.28) for the market portfolio defined in Example 3.5 for a risk-free asset with $r_{\text{f}} = 0.015$. SML can also be defined as the line that connects two special cases for $\beta = 0$ and $\beta = 1$ that correspond to $\mu_i = r_{\text{f}}$ and $\mu_i = \mu_{\text{M}}$, respectively, where μ_{M} is the expected return of the market portfolio.

model given in (3.4.13). In other words, by applying the expected value operator to both sides in (3.4.13), we get

$$E\{r_i\} = E\{\alpha_i + \beta_i r_{\mathrm{M}} + \epsilon_i\}$$

$$\mu_i = \alpha_i + \beta_i \mu_{\mathrm{M}}.$$

If the *alpha* of the *i*th asset is very high or very low, i.e., $\alpha_i \gg 0$ or $\alpha_i \ll 0$, we say *i*th asset *out-performs* or *under-performs* the market, respectively. For example, a technology company can introduce a game-changer product that would let the company and its stock out-perform their peers in the same industry for a period of time. Assets that are located above or below the SML are referred to as *over-valued* and *under-valued* since they provide higher and lower expected returns, respectively, for a given systematic risk.

3.5 RELATIVE VALUE AND FACTOR MODELS

In many cases, value (in our context, price) of an asset "tracks" the value of another asset. For example, it is reasonable to expect that stock price of a company that manufactures video processing chips is affected by the stock price of another company that is a major buyer of the chips, say a TV manufacturer.

In the previous section, we discuss in detail a special case of the relative value model, i.e., CAPM, that explains the return of an asset "relative to" the returns of the market portfolio and a risk-free asset. In this section, we start with generalizing the concept for the case of two assets. Next, we extend the discussion into the multiple assets. Then, we discuss the *factor model* which is the further generalized version of the relative value model. Finally, we discuss the *eigenportfolios*, as a type of factor model. Eigenportfolios help us to present a framework for the statistical arbitrage trading strategy (Section 4.6).

3.5.1 Two Assets

We model the return of an asset in time relative to another asset as a function of a constant, weighted return of the other asset, and a noise as given

$$r_1(n) = \alpha + \beta r_2(n) + \epsilon(n), \tag{3.5.1}$$

where α, β, and $\epsilon(n)$ are called the drift, the systematic component, and the idiosyncratic component, respectively. Choosing the asset used to model the returns, i.e., time series $r_2(n)$, is not always a trivial task. In general, an asset within the same industry as the original asset, or an exchange traded fund (ETF) that tracks the index of the corresponding industry are used. This model is also referred to as "one factor model," as a special case for the model discussed in Section 3.5.3.

3.5.2 Multiple Assets

A generalization of the model given in (3.5.1) is made where return of the jth asset is expressed as a weighted sum of multiple assets and a prediction noise term as follows

$$r_j(n) = \sum_{i=1}^{N} \beta_{j,i} r_i(n) + \xi_j(n), \tag{3.5.2}$$

For simplicity, drift and idiosyncratic component in (3.5.1), α and $\epsilon(n)$, respectively, are embedded into the residual return, i.e., $\xi(n) = \alpha + \epsilon$. In practice, the number of samples for historical asset returns is limited. Assuming that there are M samples each for N assets, (3.5.2) can be written in the matrix form as

$$\mathbf{r}_j = \mathbf{R}\boldsymbol{\beta}_j + \boldsymbol{\xi}_j, \tag{3.5.3}$$

where $\mathbf{r}_j$ is the $M \times 1$ return vector for the jth asset, $\mathbf{R}$ is the $M \times N$ matrix of asset returns, $\boldsymbol{\xi}_j$ is the $M \times 1$ idiosyncratic component vector, and $\boldsymbol{\beta}_j$ is the $N \times 1$ systematic component vector. This model is also referred to as "multiple factor model," as a special case for the model discussed in the next section. Regression coefficient vector $\boldsymbol{\beta}_j$ given in (3.5.3) can be estimated by employing the least-squares algorithm and obtained as follows

$$\hat{\boldsymbol{\beta}}_j = \left(\mathbf{R}^{\mathrm{T}}\mathbf{R}\right)^{-1} \mathbf{R}^{\mathrm{T}}\mathbf{r}_j. \tag{3.5.4}$$

3.5.3 Factor Models

A generalization of the relative value model for multiple assets given in (3.5.2) is referred to as *factor model* and defined as

$$r_j(n) = \sum_{i=1}^{N} \beta_{j,i} f_i(n) + \xi_j(n), \tag{3.5.5}$$

where $f_i(n)$ is the ith factor and $\beta_{j,i}$ is the *factor loading* for the return of the jth asset on the ith factor. Similar to (3.5.3), the factor model in matrix form is expressed as

$$\mathbf{r}_j = \mathbf{F}\boldsymbol{\beta}_j + \boldsymbol{\xi}_j, \tag{3.5.6}$$

where $\mathbf{r}_j$ is the $M \times 1$ return vector for the jth asset, $\mathbf{F}$ is the $M \times N$ matrix of factors, $\boldsymbol{\xi}_j$ is the $M \times 1$ idiosyncratic component vector, and $\boldsymbol{\beta}_j$ is the $N \times 1$ systematic component vector. There are two assumptions in a factor model:

1. Factors and the residual returns are uncorrelated, i.e.,

$$E\left\{f_i(n)\xi_j(n)\right\} = 0 \quad \forall i,j.$$

2. Residual returns for different assets are uncorrelated, i.e.,

$$E\left\{\xi_i(n)\xi_j(n)\right\} = \delta_{ij} \quad \forall i,j.$$

where we assume all processes are zero-mean. Then, cross-correlation between returns of two assets is expressed as

$$E\left\{r_j(n)r_l(n)\right\} = \sum_{i=1}^{N}\sum_{k=1}^{N} \beta_{j,i}\beta_{l,k}E\left\{f_i(n)f_k(n)\right\}, \tag{3.5.7}$$

which is only a function of the cross-correlation between factors and the factor loadings. Compared to MPT, the number of unknowns in the factor model that needs to be estimated is drastically low, especially when N is large. However, due to the assumptions made, it is a more restricted model that may make it inconsistent with the reality. Moreover, factor selection, choosing how many and which factors, is not a straightforward problem. The answer depends on the specifics of an application. There are many types of factors used in practice including earnings of a company, unemployment rate, inflation, interest rates, statistical concepts related to the returns of other assets, and many others. See [25] and references therein for a detailed discussion on factor models. Nevertheless, it is important to choose the factors that are "signals" and not "noise" (Section 5.1). A celebrated multi-factor model is the Fama-French model that uses three factors, i.e., size of the firms, their book-to-market values, and excess return on the market [26]. Another popular factor model is the one that makes use of eigenportfolios derived from correlations of asset returns in a portfolio as discussed next.

3.5.4 Eigenportfolios

In this section, we discuss a widely used type of factor called the eigenportfolio. Let us rewrite (3.5.3) as

$$\mathbf{r}_j = \mathbf{R}\boldsymbol{\gamma}_j + \boldsymbol{\xi}_j, \tag{3.5.8}$$

where we renamed $\boldsymbol{\beta}_j$ of (3.5.3) as $\boldsymbol{\gamma}_j$. Let us decompose the correlation matrix of asset returns given in (3.2.12) into its eigenvectors and eigenvalues as follows [27]

$$\mathbf{P} = \boldsymbol{\Phi}\boldsymbol{\Lambda}\boldsymbol{\Phi}^{\mathrm{T}} \tag{3.5.9}$$

where $\boldsymbol{\Lambda}$ is a diagonal matrix with the eigenvalues as its elements expressed as

$$\boldsymbol{\Lambda} = \begin{bmatrix} \lambda_1 & . & . & 0 \\ 0 & \lambda_2 & . & 0 \\ & & \ddots & \\ . & . & & . \\ 0 & . & . & \lambda_N \end{bmatrix},$$

λ_k is the kth eigenvalue with $\lambda_k \geq \lambda_{k+1}$, $\lambda_k \geq 0\ \forall_k$, $\sum_k \lambda_k = N$, $\boldsymbol{\Phi}$ is an $N \times N$ matrix comprised of N eigenvectors as its columns

$$\boldsymbol{\Phi} = \begin{bmatrix} \boldsymbol{\phi}_1 & \boldsymbol{\phi}_2 & \cdots & \boldsymbol{\phi}_N \end{bmatrix}, \tag{3.5.10}$$

and $\boldsymbol{\phi}_k$ is the $N \times 1$ eigenvector corresponding to the kth eigenvalue. Moreover, $\boldsymbol{\Phi}\boldsymbol{\Phi}^{\mathrm{T}} = \mathbf{I}$ due to the orthonormality property of the eigenvectors [27]. Given the volatility matrix defined in (3.2.9) is diagonal, i.e., $\boldsymbol{\Sigma}^{-1}\boldsymbol{\Sigma} = \mathbf{I}$, (3.5.8) can be rewritten as

$$\begin{aligned} \mathbf{r}_j &= \mathbf{R}\boldsymbol{\Sigma}^{-1}\boldsymbol{\Phi}\boldsymbol{\Phi}^{\mathrm{T}}\boldsymbol{\Sigma}\boldsymbol{\gamma}_j + \boldsymbol{\xi}_j \\ &= \mathbf{F}\boldsymbol{\beta}_j + \boldsymbol{\xi}_j, \end{aligned} \tag{3.5.11}$$

where $\mathbf{F} \triangleq \mathbf{R}\boldsymbol{\Sigma}^{-1}\boldsymbol{\Phi}$ is the $M \times N$ principal components matrix [28] with its elements $f_k(n)$, $n = 0, 1, \ldots, M-1$ being the nth sample value of the kth principal component which is given as

$$f_k(n) = \sum_{i=1}^{N} \frac{1}{\sigma_i} r_i(n)\phi_i^{(k)}. \tag{3.5.12}$$

where $\phi_i^{(k)}$ is the ith element of the kth eigenvector, $\boldsymbol{\phi}_k$, and $\boldsymbol{\beta}_j \triangleq \boldsymbol{\Phi}^{\mathrm{T}}\boldsymbol{\Sigma}\boldsymbol{\gamma}_j$. Note that (3.5.12) is merely a weighted sum of returns for an N-asset portfolio. Since weights are functions of the eigenvectors, principal component given in (3.5.12) is also called *eigenportfolio* [29]. Moreover, (3.5.11) is nothing else but (3.5.6) in which eigenportfolios are the factors. There are three immediate advantages of using eigenportfolios as factors. Namely,

1. Cross-correlation between two eigenportfolios is zero, and the variance of a particular eigenportfolio is equal to its corresponding eigenvalue, i.e.,

$$E\{f_i(n)f_j(n)\} = \begin{cases} \lambda_i & i=j \\ 0 & i \neq j \end{cases}. \tag{3.5.13}$$

   Therefore, number of unknowns in (3.5.7) is significantly reduced.
2. Given $\lambda_k \geq \lambda_{k+1}$, it is practical (and advantageous as discussed in Section 5.1) to use only the first $L \ll N$ eigenportfolios in (3.5.11). This practice naturally filters the noise in the empirical correlation matrix of measured asset returns in a portfolio (3.2.12).
3. Eigenportfolios are tradable as they are portfolios of tradable assets.

We revisit the eigendecomposition of empirical correlation matrix of asset returns in Section 5.1 where we discuss filtering of measurement noise and its impact on portfolio risk estimation. We will revisit eigenportfolios in Section 4.6 where we discuss statistical arbitrage methods. We note that, similar to (3.5.4), regression coefficient vector $\boldsymbol{\beta}_j$ given in (3.5.11) can be estimated by using the least-squares method and expressed as

$$\hat{\boldsymbol{\beta}}_j = \left(\mathbf{F}^{\mathrm{T}}\mathbf{F}\right)^{-1}\mathbf{F}^{\mathrm{T}}\mathbf{r}_j. \tag{3.5.14}$$

3.6 SUMMARY

Geometric Brownian motion is a widely used mathematical model for asset prices with the assumption of their constant volatilities. There are more sophisticated price models such as the Heston model that incorporate the variations of asset volatility. Return of an asset, its second order statistics (expected return and the volatility), pairwise cross-correlations of asset returns for a portfolio are commonly used statistics in financial models for investment decisions. Jump, an abrupt change in the return of an asset in time, is a natural and important phenomenon that needs to be considered when modeling the price of an asset. Modern portfolio theory suggests a framework to create efficient portfolios that provide minimum risk for a given expected portfolio return. Capital asset pricing model (CAPM) explains the expected return of an asset in terms of the return of a risk-free asset and the expected return of the entire market portfolio. It leads us to calculate useful mathematical tools and metrics such as the capital market line, beta of an asset, and the security market line. As a quantitative methodology, return of an asset may be expressed relative to the return of another asset, the return of an asset basket, or in general, a set of factors. Relative value models and factor models provide us with a mathematical procedure to identify the over and undervalued assets with respect to others. Eigenportfolio is a special factor model with its unique features such as pairwise perfect decorrelation of eigenportfolios and filtering of market noise in empirical correlation matrix. The topics covered in this chapter emphasize the fact that well known signal processing methods find their popular use in quantitative finance and electronic trading applications.

CHAPTER 4

Trading Strategies

Traditionally, investing in an asset means buying the asset and holding it for a prolonged time in order to profit from its future gains. For example, if we believe (or we are advised) that a company will perform good in the following year, we *invest* into that company. The easiest way of investing is to buy the company stocks. If the company does good, the stock price goes up, and our investment gains value. In a sense, *investors* (long term buyers) become partners of the company for a time period of their choice and they contribute to its growth. On the other hand, there are short term "investors" in the market who are not mainly interested in the performance of the company, but they rather speculate on price variations of its stock and stock derivatives. These players are called *traders*. They try to identify price inefficiencies and strong trends (up or down) to buy and sell stocks (not necessarily in the same order) with relatively short holding times for the objective of maximized profit at unit risk (risk normalized return).

In this chapter, our focus is on *strategies* for trading stocks. They analyze currently available market data and information based on a predefined framework to detect indicators of price inefficiencies or market trends that lead to open positions in such stocks. A trading strategy may be *ad-hoc* or *systematic*. In an ad-hoc strategy, one may detect and act upon specific events in the market rather than trading according to a set of rules created

A Primer for Financial Engineering. http://dx.doi.org/10.1016/B978-0-12-801561-2.00004-6

through extensive modeling and backtesting. In contrast, systematic trading is the methodical way to trade according to predefined and well-tested rules, controls, and schedules. The systematic traders continuously test the strategy and recalibrate its parameters.

Trading strategies commonly utilize signal processing methods such as eigenanalysis [27, 30–32], wavelets [33, 34], machine learning [35], neural networks [36, 37], hidden Markov models [38], evolutionary algorithms [39], and many others. Most commonly used trading strategies may be grouped under two categories: *mean reversion* and *trend following*. We delve into the first through pairs trading and statistical arbitrage in Sections 4.5 and 4.6, respectively, and discuss the latter in Section 4.7. At the end of each section, we provide recipes that summarize the significant steps of the trading strategy. We also provide the MATLAB implementations of the strategies for the readers of further interest. We focus on stocks in the chapter since the strategies and models we discuss are originally developed mostly for trading stocks. However, similar concepts may also be applicable or relevant to trade other financial instruments.

We can also group the strategies, and traders who trade with them, into three categories: *fundamental*, *technical*, and *quantitative traders* (also know as *quants*). Fundamental traders evaluate the financials (cash flow, earnings, price-earning ratios, etc.), business performance, credit risk and other performance metrics of a company to make trading decisions for that stock. On the other extreme, technical traders just focus on the price. They use various charting techniques (slop lines, supports, trends, etc.) to make trading decisions on the stock. A relatively new breed of traders, quantitative traders, look for inefficiencies in the market through studying and developing complex mathematical models and statistical analysis methods to make trading decisions. Quants bring in their strengths in mathematics, physics, and engineering to develop quite sophisticated trading strategies and their execution through computers in real time. In this chapter, our focus is mainly on quantitative trading strategies.

Trading strategies discussed in this chapter share two common phases. They are *analysis* and *signal generation* for trades. In the former, we analyze the raw market data and look for indicators to identify relative price inefficiencies between a pair of stocks (Section 4.5), to highlight arbitrage opportunities across industries (Section 4.6), or to predict strong upward or downward trends in the price (Section 4.7). In the latter, we *decide* on trading a certain stock (generate signals to open or close positions) based on the indicators generated in the first phase. For example, a question like "which method should I use to detect if there is a trend?" falls under the analysis phase whereas the question "when I detect a trend, what rules should I use to decide on trading a particular stock (opening a position)?" falls under the signal generation phase.

4.1 TRADING TERMINOLOGY

We overload the word "signal" in this chapter since it has two distinct meanings in signal processing and financial trading literatures. In the former, a signal is a deterministic or random function that conveys information, such as price, volume, colored noise, and others. In the latter, a signal is an indicator to trade a stock (open or close a position in the stock) such as buy, sell, short-sell, buy-to-cover, etc. We delve into details of trading signals and positions in Section 4.2.

In traditional portfolio management (Chapter 3), we rebalance a portfolio, move from one investment allocation vector (3.3.5) to another, whenever the underlying fundamentals or statistics change. In some cases, *continuous rebalancing* is needed in which the number of units we own in each stock in the portfolio is readjusted at every incoming data sample. However, in the trading strategies discussed in this chapter and risk management methods discussed in Chapter 5, we only change our position in a stock when an enter or exit signal is generated based on the predefined rules of the strategy.

The number of stocks we use to perform analysis can be (and usually is) different than the number of stocks we trade. For example, one can perform analysis on all the technology stocks, but may choose to trade only a small set (*basket*) of stocks. Since they are distinct set of stocks, we use different names for the two. The former is called the *trading universe* and the latter is the *portfolio*.

Today, computers are used for almost all systematic trading. The strategies we discuss boil down to lines of code in programming language of choice for one reason or another. Therefore, one might think that algorithmic trading, also called program trading, is essentially software implementation of a trading strategy. On the contrary, the term *algorithmic execution* is used for getting in and out of a position (usually a very large one), i.e., executing the requested trade by a portfolio manager or trader, according to a schedule and other trade execution parameters. We discuss algorithmic trading and execution strategies in Section 6.1.

4.2 LONG AND SHORT POSITIONS

Buying or selling shares of a stock puts the trader in a *position* in that stock. This position has a market value varying in time. Trader is in *long* or *short* position if the shares he or she owns is positive or negative, respectively. It is said that the trader is *not in a position* when the number of owned shares is zero. We create (open) a position of a stock at discrete-time n_0 through a trade. Similarly, an open position is closed through the complementary trade at $n_1 > n_0$. The return on the investment (open position for a period of time like hour, day, week) through the life of the trade (buy first-then-sell or sell first-then-buy a stock) at discrete time n is given as

$$r_{\text{inv}}(n) = q(n-1)r(n), \tag{4.2.1}$$

where $r(n)$ is the rate of return of the stock defined as

$$r(n) = \frac{p(n) - p(n-1)}{p(n-1)}, \tag{4.2.2}$$

$p(n)$ is the price of the stock at time n, and $q(n-1)$ is the amount of capital, e.g., in US dollars, invested in that stock at discrete-time $n-1$. We note that

$$q(n) = x(n)p(n), \tag{4.2.3}$$

where $x(n)$ is the number of shares we own in that stock, and $x(n) = x(n-1)$ as long as there is no trading signal at discrete time n. From (4.2.1), (4.2.2), and (4.2.3), we have

$$r_{\text{inv}}(n) = x(n)\left[p(n) - p(n-1)\right].$$

When trader is in a long position, $x(n) > 0$, $r_{\text{inv}}(n)$ can be only positive for $r(n) > 0$. In other words, a long position can only profit when the price of the stock goes up. Similarly, in a short position where $x(n) < 0$, $r_{\text{inv}}(n)$ is only positive when $r(n) < 0$. In other words, a short position generates profit when the price of the stock goes down during the *holding time*. A profitable trading strategy is the one that makes good judgment on when to open and close a long or short position such that the return on investment is positive.

Bull and *bear markets* are the ones in which market prices of the stocks, in general, trend up and down, respectively, for a prolonged time. Therefore, when a trader is in long or short position in a stock, it is said that the trader has *bullish* or *bearish position* in the stock, respectively.

In order to get in and out of a long position, a trader must *buy* and *sell* a stock, respectively. On the contrary, in order to get in and out of a short position, a trader must *sell* and *buy* a stock, respectively. In order to short a stock with the expectation of future price decline, the trader must first borrow the shares from a *lender* with a predefined *fee* (rent) and sell them at the current market price. Then, when the trader wants to get out of the short position, he needs to buy the shares back from the market and return them to the lender along with fee payment. If the price of the stock goes down during the holding time of a *short position*, trader sells high and buys low, and makes profit. The difference between a long position and short position is the order of buying and selling a stock. The terms such as *selling short* and *buying to cover* are also used to describe *these moves related to short* selling. It is common to differentiate the four moves as *buying to open*, *selling to close*, *selling to open*, and *buying to close* where terms *open* and *close* mean *getting in* and *getting out* of a position, respectively. The execution of *short selling* is less complicated than it may sound. It is commonly the *broker-dealer* (BD) that handles the borrowing and returning of the shares for the traders. Therefore, from trader's perspective, the mechanics of getting in a long or short position is essentially the same. However, short selling a stock might not be possible all the time. Several such situations are summarized as follows.

1. Short selling is banned by law in some countries.
2. Short selling may be restricted for stocks with steep decline from the previous day's close price.
3. During the initial public offering (IPO) of the stocks, it is usually not allowed to short them for a period of time.
4. The broker temporarily might not be able to borrow that particular stock due to shortage of available shares.

In general, *less liquid* stocks are harder to borrow and short sell. Those stocks are referred to as *hard-to-borrow* or *non-shortable*. Moreover, short selling may be riskier than long positions as highlighted below.

1. Trader pays interest on the cash received due to short selling of the borrowed shares. However, many brokers handle exchange of stocks due to short selling among their customers internally, and do not charge interest on short sells for small orders. This should not be confused with the term *short interest* that describes the total number of company shares currently sold short.
2. If the stock is illiquid and there are not always shares immediately available in the market, it might be hard to get out of the short position. Trader may be rejected by the BD with *"we are unable to locate the shares to cover the position."*

3. The worst that can happen when one is in a long position when the price goes down very close to zero. Therefore, theoretical loss has a limit when one is in a long position. However, in a short position, theoretically, the loss is infinite since there is no upper limit on a stock price.
4. Shorting requires opening a *margin account* with BD where trader does not earn interest. If the loss due to a short position gets too high, the broker can make a *margin call.* In this case, trader needs to inject additional cash into the margin account or the broker will liquidate the longed stocks. Margin calls may tighten cash flow and disturb a running trading strategy.

Some people consider short selling as immoral. They believe that shorting a stock is nothing else but speculation. It hurts the company issuing the stock, and it prepares the ground for a bear market. Hence, bad economy. However, it is argued in [40] that restrictions on short selling may result in increased volatility in the market.

4.3 COST OF TRADING

Buying and selling stocks are not for free. Every order we place to an exchange through a broker incurs a *transaction cost.* There is always *slippage* between entry (or exit) price of a desired trade and its executed price. The slippage may be significant for certain *illiquid stocks*. Transaction costs can be fixed or can change depending on the size of the order. For example, a broker may charge us $0.05 per share on the buy or sell of a stock, but they may also impose minimum transaction costs, say minimum $1 per transaction. In some markets and countries, trading cost is not defined as per traded share but in percentage of the dollar value of the order. In the former, it is cheaper to trade pricy stocks since one trades less number of units for the same dollar value. Brokers usually offer tiered pricing in terms of volume such that lower fees apply to those clients who place large numbers of orders. Some brokers charge lower fees to those customers who trade at high frequencies (Section 6.4).

Type of order affects the trading cost. *Limit* and *market orders* we discuss in Section 6.2.1 have different trading cost since they add and remove liquidity from the market. Limit orders are rewarded to attract more liquidity to the market. ECN fees are usually combined in broker fees for a real life trading scenario. Some brokers provide additional order types (algorithmic execution types) that guarantee certain goals like the order to be executed at the mid price (the average of the best bid and best ask prices) for additional fee.

Historical and real-time market data play a crucial role in the development, backtesting, and running of trading algorithms. Most trading models utilize historical data in their analysis and decision processes. Therefore, broker-dealers and other financial data companies provide near real-time market data flows to their customers bundled with their services. *End of day (EOD)* historical price data for most US stocks can also be downloaded from Yahoo! Finance[1] for free. However, depending on how sophisticated the trading strategy is, one might need *direct market access (DMA)* in particular for high frequency real-time market data from exchanges or data providers. These services also add to the cost of trading.

Example 4.1. Assume that we want to short sell a stock. Broker charges \$0.005/share for the order. Interest rate on the cash for selling borrowed shares of this particular stock is 0.005% each overnight. ECN charges \$0.0035/share for the market-orders. We place a sell to open (Section 4.2) market order to short sell 300 shares. Our order is executed at an average price of \$78.412/share. Broker charges \$0.005/share × 300 shares = \$1.5. ECN charges \$0.0035/share × 300 shares = \$1.05. We place a buy to cover market order *next day* to exit the short position. Overnight interest on the cash is \$78.412/share × 300 shares × 0.005% = \$1.18. Our order is executed at an average price of \$78.08. Similarly, broker and ECN charges are \$1.5 and \$1.05 for the exit order, respectively. Hence, *total trading cost* of this finished trade is \$6.28. The gross profit on the trade is (\$78.412 – \$78.08) /share × 300 shares = \$99.60. Therefore, the cost of trading is 6.3% of the gross profit in this case.

4.4 BACKTESTING

The performance of a trading strategy designed and implemented for algorithmic trading using currently available historical market data is thoroughly tested and considered as an important indicator to go live and trade with it. This crucial step is called *backtesting*. It is also used to calibrate the

---

[1] http://finance.yahoo.com/. See corresponding terms of use as well.

parameters of a strategy. Typically, backtesting process consists of the following steps.

1. Acquire the historical market data from a reliable source (price and trading volume of the stocks in our trading universe, and other relevant financial data considered in the strategy),
2. Run the analysis phase of the strategy to generate trading signals,
3. Generate the profit and loss and measure its performance,
4. Calibrate the parameters like thresholds and time windows for statistical measurements, and retest the strategy.

Forward testing (also known as *paper trading*) is another form of evaluating a trading strategy where live (real-time) market data is used rather than market simulator utilizing currently available historical data. Forward testing may provide additional information on the performance of the strategy. We emphasize that backtesting can be performed within seconds or minutes since the historical data is already available. In contrast, forward testing takes much longer as it uses live market data in real-time for the desired test duration.

4.4.1 Profit and Loss of a Trading Strategy

The most common figure of merit used in backtesting is the performance of the portfolio equity, $E(n)$, called the *profit and loss (P&L)*. P&L is an equation in $E(n)$. Since the first thing we do with the realization of a P&L equation (through backtesting or forward testing) is to plot it, sometimes we also call it the *P&L curve*. P&L equation defines how the portfolio equity evolves in time. Portfolio equity is usually measured in the form of a currency like US dollars. Interest on cash, our positions in the stocks with their changing market values, and the transaction costs affect the portfolio equity. Let us start with the interest rate. It is possible that we might not have an open position in any of the stocks at some point in time. In that case, we earn only the interest according to the following P&L equation:

$$E(n) = E(n-1) + E(n-1)r_{\text{f}}, \tag{4.4.1}$$

where r_{f} is the interest rate and n is the time index for the trading period. When we are in a position for the ith stock in the portfolio, the P&L equation is written as follows:

$$\begin{aligned} E(n) &= E(n-1) + E(n-1)r_{\text{f}} \\ &\quad + q_i(n-1)r_i(n) - q_i(n-1)r_{\text{f}} \end{aligned} \tag{4.4.2}$$

where $q_i(n-1)$ is the value of our position defined in (4.2.3) and $r_i(n)$ is the return of the stock defined in (4.2.2) at discrete time n. According to the third term in (4.4.2), one profits or loses on investment in a stock depending on its return during the *holding time*. Last term in (4.4.2) describes interest we receive (pay) on the amount we invest for short (long) positions (cost of invested capital). In other words, we can use the cash received by selling the borrowed stocks when shorting to earn interest. As discussed in Section 4.3, we need to account for the *cost of trade*. We add cost term to the P&L equation as

$$\begin{aligned} E(n) = E(n-1) &+ E(n-1)r_f \\ &+ q_i(n-1)r_i(n) - q_i(n-1)r_f \\ &+ |q_i(n) - q_i(n-1)|\, I_i(n)\epsilon, \end{aligned} \tag{4.4.3}$$

where ϵ is the cost of trade in percentage, $|\cdot|$ is the absolute value operator, and $I_i(n) \to \{0, 1\}$ is the *indicator function* with the value of 1 when there is a transaction (buying or selling) on the ith stock, and 0 otherwise at time n. Now, we define the P&L equation for a portfolio of investment instruments like US equities for a trading strategy as follows [29, 41].

$$\begin{aligned} E(n) = E(n-1) + E(n-1)r_f + \sum_{i=1}^{N} q_i(n-1)r_i(n) \\ - \sum_{i=1}^{N} q_i(n-1)r_f - \sum_{i=1}^{N} |q_i(n) - q_i(n-1)|\, I_i(n)\epsilon, \end{aligned} \tag{4.4.4}$$

that is similar to (4.4.3) summed over all the stocks in the portfolio. When we backtest a trading strategy, we usually start with $E(0) = \$100$ and $q_i(0) = 0\ \forall i$, and evolve the value of our position on the ith stock in time n as

$$q_i(n) = \begin{cases} \pm\delta E(n) & I_i(n) = 1 \text{ and } q_i(n-1) = 0 \\ 0 & I_i(n) = 1 \text{ and } |q_i(n-1)| > 0\,, \\ q_i(n-1)\,[1 + r_i(n)] & \text{otherwise} \end{cases} \tag{4.4.5}$$

where $0 < \delta \leq 1$ is the *capital allocation ratio*. It is a real number that describes the percentage of the total portfolio equity we allow trading (opening a position) for each stock triggered by each enter signal generated by the strategy (Section 4.4.4).

See files `backtesting.m`, `equity.m`, and `invest.m` for the MATLAB implementations of a backtester, portfolio equity calculator (according to (4.4.4)), and investment calculator (according to (4.4.5)), respectively.

4.4.2 Performance Measures

We measure and describe the performance of a trading strategy by calculating average return and the volatility (risk) of its portfolio equity, given in (4.4.4), as follows:

$$\mu_E = E\{r_E(n)\}$$
$$\sigma_E = \left(E\left\{r_E^2(n)\right\} - \mu_E^2\right)^{1/2}, \tag{4.4.6}$$

respectively, where the normalized return $r_E(n)$ is calculated as

$$r_E(n) = \frac{E(n) - E(n-1)}{E(n-1)}. \tag{4.4.7}$$

In any trading strategy, one aims to have a portfolio equity with high average return and low volatility. Similar to the single stock case (Section 3.2), the Sharpe ratio of a portfolio equity, *risk-adjusted return*, is defined as follows:

$$\mathrm{SR} = \frac{\mu_E - r_f}{\sigma_E}. \tag{4.4.8}$$

Sharpe ratio quantifies excess return of a trading strategy over risk-free return for a unit risk. The higher the Sharpe ratio the better the strategy is. It is customary to report *annualized* Sharpe ratio. If the portfolio equity is calculated daily, then Sharpe ratio calculated with (4.4.8) is called the *daily* Sharpe ratio. Annualized Sharpe ratio is calculated by multiplying the daily mean and daily variance of the normalized return, $r_E(n)$, with the number of days in a year, N, as given

$$\mathrm{SR}_{\mathrm{annualized}} = \frac{(\mu_R - r_f) \times N}{\sigma_E \times \sqrt{N}} = \mathrm{SR} \times \sqrt{N}.$$

Note that there are 252 days in the United States in which the stock markets are open, hence in this case $N = 252$. Another performance measure that we derive from the P&L curve is the *maximum drawdown*. It is the maximum continuous decline from the peak of the P&L during a time window (usually the entire time frame used in backtesting), and defined as

$$\mathrm{MD} = \max_{n \in [0,N)} \left[\max_{k \in (0,n)} E(k) - E(n) \right].$$

The quantities of interest are how long (the time interval) and how low (the continuous loss) the maximum drawdown is. A good trading strategy is the one without any drastic drawdowns and a low maximum drawdown.

See files `sharpe_ratio.m` and `max_drawdown.m` for MATLAB implementations of Sharpe ratio and maximum drawdown calculations, respectively.

4.4.3 Backtesting a Trading Strategy

We note that the P&L equation defined in (4.4.4) does not take market impact into account. We discuss market impact in detail in Section 6.1. Moreover, the model of trading cost in (4.4.4) is simplistic as it assumes that lending and borrowing costs are the same. A better trading cost model would be the one that takes into account the differences between *shorting* and *longing* a stock as well as *limit* and *market orders* in terms of cost as we discuss in Section 4.3. The P&L equation we provide in (4.4.4) is a good approximation for academic studies. However, in order to have a better fit with reality, a thorough backtesting of a trading strategy must involve good *market impact* and *cost of trading* models as discussed in Sections 6.1 and 4.3, respectively. Moreover, though it is not straightforward to implement, incorporation of a *limit order book (LOB) model* (Section 6.2.4) in a backtester would potentially make it a far better performance simulator for a given trading strategy.

4.4.4 Leverage

Traditionally, it is thought that one can invest in stocks up to the value of their portfolio equity. For example, it is commonly the belief that one can invest \$1,000 in the stock market against the cash in his or her bank account. However, in reality, traders can invest in stocks that sum up to a larger value than their equities in their investment accounts. The ratio between the former and the latter is called *leverage* (also known as *gearing*) and defined as

$$L(n) = \frac{1}{E(n)} \sum_{i=1}^{N} |q_i(n)|, \tag{4.4.9}$$

where $q_i(n)$ is the current value of the ith stock (4.4.5), $E(n)$ is the current value of portfolio (4.4.4), and N is the number of stocks in the portfolio. Leverage is commonly offered by financial institutions like broker-dealers (BDs). For example, if one buys \$1000 worth of stocks with the available capital of \$1000, the leverage of that investment is $L = 1$. In USA,

Regulation T (also known as Reg-T)[2] limits the maximum leverage for overnight positions of retail traders to be $L \leq 2$. However, broker-dealers can offer significantly higher leverage limits (typically $L \leq 4$) for intraday trades (those that are opened and closed within the market hours in the same day without any overnight position).

Example 4.2. A trader in the stock market has a long position of 40 shares in Apple Inc. (APPL) and a short position of 156 shares in Cisco Systems, Inc. (CSCO). Their current market prices are \$98.97 and \$25.61 per share, respectively. Trader has put a start up investment capital of \$5000 in the trading account. The leverage for this trivial portfolio is calculated from (4.4.9) as

$$\begin{aligned} L(n) &= \frac{1}{\$5000}\left[|\$98.97 \times 40| + |\$25.61 \times (-156)|\right] \\ &\cong 1.59. \end{aligned}$$

One always monitors and controls the leverage limits in a trading strategy and its backtesting by adjusting the capital allocation ratio for a single stock, δ, as described in (4.4.5). In other words, the higher the capital allocation ratio δ, the higher the leverage L. It is observed from (4.4.4) and (4.4.5) that the average return and volatility of the portfolio equity defined in (4.4.6) hence the Sharpe ratio defined in (4.4.8) are functions of investment ratio. However, if the interest rate, r_{f}, is negligible compared to the average return, μ_{E}, investment ratio inherent in both the numerator and denominator and the Sharpe ratio is canceled. Therefore, increasing the investment ratio, δ, hence the leverage, does not affect the Sharpe ratio of a strategy. In other words, increasing leverage will increase not only the average return, μ_{E}, but also the volatility, σ_{E}.

4.5 PAIRS TRADING AND MEAN REVERSION

Pairs trading was discovered by Gerry Bamberger and Nunzio Tartaglia at Morgan Stanley in the early 1980s. It is still used by some traders, and it is also the predecessor of many more complex trading strategies developed

[2] 12 CFR 220 – Code of Federal Regulations, Title 12, Chapter II, Subchapter A, Part 220 (Credit by Brokers and Dealers).

afterwards. We discuss the statistical arbitrage as one of those strategies in the next section.

The idea of pairs trading is simple. A *pair* of two correlated stocks, i.e., their historical prices, $p_1(n)$ and $p_2(n)$ pretty much *trace* each other, are traded together. This is achieved by picking two stocks from the same industry, such as American Airlines and Continental Airlines, or Coca-Cola and Pepsi Co. Then, their price *spread* at time n is calculated as

$$\Delta(n) \triangleq p_1(n) - p_2(n). \tag{4.5.1}$$

Once we have the spread, the trading strategy is the following. At any point in time, if spread gets *too high* or *too low* compared to its historically measured value, then, *bet against the spread.* In other words, *short sell* the first stock and *buy* the second one (essentially, *short sell the spread*). When spread returns back to its historical mean, *buy* the first stock *to cover* the short position and *sell* the second stock. Similarly, make the opposite move, i.e., *buy the spread,* if spread gets too low compared to its historically measured value and *sell the spread* when it is back to its historical mean.

When spread is too high (or low), it is expected that it will get lower (or higher), in the near future. In other words, it is expected that it will *revert back to its mean*. It is assumed that spread is a *mean-reverting* signal. Therefore, trader expects to profit from the short (long) position they have in the spread. While in the position, if the prices of both stocks trend up or down, trader will be immune to this trend by being both long and short in two stocks at the same time. Hence, bet stays neutral to these price movements. The short-lived deviations in the spread between the prices of the two historically correlated stocks are usually alluded to market over-reaction. For example, traders in the market may over-react to news about one of the companies, and they *buy* or *sell* the stocks aggressively. This would naturally drive the price of the stock *up* or *down*, respectively. Such a price change could eventually drive the spread away from its historical mean, accordingly. Two scenarios we covered here are examples of an *arbitrage*, i.e., a trading opportunity for the trader to take advantage of price inefficiency in the market.

4.5.1 Model Based Pairs Trading

Relative value model discussed in Section 3.5, specifically model for the two stock scenario given in (3.5.1), can be used toward developing a pairs trading strategy. According to the model, we have the following relationship between the returns of two stocks, $r_1(n)$ for the first stock and $r_2(n)$ for the second stock

$$r_1(n) = \alpha + \beta r_2(n) + \epsilon(n), \tag{4.5.2}$$

where α, β, and $\epsilon(n)$ are the drift, systematic component, and idiosyncratic component, respectively. Ignoring the transaction costs and interest on cash for the time being due to simplicity, let us express the return on investment, $r_{\text{inv}}(n)$, for investing 1 dollar (or any other form of currency) in the first stock and $-\beta$ dollars in the second stock (short sell the second stock) of the pair as

$$r_{\text{inv}}(n) = r_1(n) + (-\beta)\, r_2(n). \tag{4.5.3}$$

We note that, this is equivalent to setting the *investment allocation vector* of (3.3.5) to

$$\mathbf{q} = \begin{bmatrix} 1 \\ -\beta \end{bmatrix}.$$

Substitution of (4.5.2) into return on investment (4.5.3) yields

$$r_{\text{inv}}(n) = \alpha + \epsilon(n). \tag{4.5.4}$$

Expected return on investment is calculated as

$$\mu_{\text{inv}} = E\{r_{\text{inv}}(n)\} = \alpha + E\{\epsilon(n)\}. \tag{4.5.5}$$

Similarly, variance of the return on investment is calculated as

$$\text{var}\,[r_{\text{inv}}(n)] = \text{var}\,[\alpha + \epsilon(n)] = \text{var}\,[\epsilon(n)],$$

since α is a constant. Therefore, the risk of the investment involves only the idiosyncratic component, $\epsilon(n)$. In other words, the risk is not the function of β. Expected return of the investment given in (4.5.5) suggests two trading strategies as explained below.

1. The drift is large, $\alpha \gg 0$. This means that the first stock historically has outperformed the stock and it is expected to continue to do so. In this case, this investment is profitable over a prolonged period. This scenario is similar to the one we investigated in Section 3.4.6 for discussing the SML. In that case the second stock simply replaced with the market portfolio.
2. The drift is relatively small in the time window under consideration, $\alpha \cong 0$. However, the recent samples of the idiosyncratic component, $\epsilon(n)$, are abnormally large negative values. In other words, although historically both stock prices move together where samples of $\epsilon(n)$ are

small, the second stock has outperformed the first stock in a relatively short time period. However, statistically we *expect* this anomaly to recover (we expect more positive samples than negative samples of $\epsilon(n)$ will be observed in the near future). This recovery is profitable since return on investment (4.5.4) is $r_{\text{inv}}(n) = \epsilon(n)$ in such a scenario.

Let us further elaborate the idiosyncratic component, $\epsilon(n)$, in order to understand why the second trading strategy is of the actual interest for pairs trading. The log-returns of stocks, $g_1(n)$ and $g_2(n)$, from (3.2.13), can be written as

$$s_i(n) = s_i(0) + \sum_{j=1}^{n} s_i(j), \tag{4.5.6}$$

where $s_i(n) = \ln[p_i(n)]$ is the log-price of the ith stock. In other words, log-price of the ith stock at discrete-time n is the sum of its log-price at $n = 0$ and all the log-returns in between. We already know from Section 3.2 that returns are approximately identical to the log-returns. Substitution of (4.5.2) in (4.5.6) and evaluating for the first stock yields

$$\begin{aligned} s_1(n) &= s_1(0) + \sum_{j=1}^{n} r_1(j) \\ &= s_1(0) + \sum_{j=1}^{n} [\alpha + \beta r_2(j) + \epsilon(j)] \\ &= s_1(0) + n\alpha + \beta \sum_{j=1}^{n} r_2(j) + \sum_{j=1}^{n} \epsilon(j). \end{aligned} \tag{4.5.7}$$

Similarly, we have $s_2(n)$ for the second stock as

$$s_2(n) = s_2(0) + \sum_{j=1}^{n} r_2(j). \tag{4.5.8}$$

Since we assume that the drift between the returns of two stocks, α, is almost zero, within the observation window, $\alpha \cong 0$, from (4.5.7) and (4.5.8) we have

$$\begin{aligned} \sum_{j=1}^{n} \epsilon(j) &= s_1(n) - s_1(0) - \beta\,[s_2(n) - s_2(0)] \\ &= \ln\left[\frac{p_1(n)}{p_1(0)}\right] - \beta \ln\left[\frac{p_2(n)}{p_2(0)}\right]. \end{aligned} \tag{4.5.9}$$

This equation is nothing else but the spread between two stocks, similar to the basic spread definition given in (4.5.1). The only difference is the fact that we monitor the beta value of the stock, normalize to the initial price, and operate on the log-price instead of the price. Hence, the idiosyncratic component, $\epsilon(n)$, is the return of the spread. Therefore, expecting spread to return from a lower value back to its mean leads one to expect observing higher number of positive samples than negative samples of $\epsilon(n)$ in the near future.

In summary, spread is the cumulative sum of the error terms, $\epsilon(n)$. In pairs trading, traders make their trade decisions based on the spread and its statistics (such as mean and variance). It is common practice to use a model for the spread itself and estimate the statistics according to the model. Since we know spread is (or should be) a mean-reverting process, it makes sense to use such a model. Among many different mean-reverting process models available in the literature, the most commonly used two spread models are as follows.

1. Auto-regressive (AR) process [27, 42], more specifically Ornstein-Uhlenbeck process (O-U process) [29, 43].
2. Fractional geometric Brownian motion process [44] and their variations (see for example [45] and references therein).

Up to this point, we have only discussed the case where spread is lower than its historical mean and expected to go back to it. In the opposite case where the first stock is over-priced, the logical move is not to get in long but rather a short position in the pair. It means to invest -1 dollar (short) in the first stock and β dollars (long) in the second stock. We also note that we ignored transaction costs in (4.5.3). However, in making a decision to open a position into a spread, one needs to make sure that the spread is large enough to cover the transaction costs.

In contrast to trend following (momentum based) strategies where only one stock is traded in a trade, in pairs trading, a trade naturally involves two stocks to be traded at the same time. In a sense, we are investing in the first stock and also protecting ourselves from a potential loss by simultaneously investing in the second stock. Hence, we are *hedging* our investment in the first stock with the simultaneous investment in the second stock.

4.5.2 Market Neutrality

Instead of trading a stock against another stock, we can also trade a stock against a factor (assuming the factor is tradable). A factor can be the market portfolio (Section 3.4) or it may be an ETF (Section 2.2.4). The return on the investment for such a case is expressed as

$$r_{\text{inv}}(n) = r_1(n) + (-\beta)\, r_{\text{M}}(n),$$

where $r_{\text{M}}(n)$ is the return of the market portfolio or an ETF that mimics the market. The latter is usually more efficient, since one would need to trade all the stocks in the market portfolio in order to get in and out of a trade when market portfolio is the factor. This would incur a large transaction cost. However, in the ETF case, one still needs to trade only two financial instruments incurring much lower transaction costs. For example, the stock could be APPL and a proper ETF may be QQQ that tracks the NASDAQ 100 index.

In the general case, the expected return on investment is not a function of the β value, the beta of the stock to the *market,* and identical to (4.5.5). The trading strategies where expected return is not a function of the market are referred to as *market-neutral.* The readers with more interest may find a detailed discussion on trading stocks against ETFs in [29]. In such a case, we hedge our trade with the market portfolio or the ETF.

4.5.3 A Recipe for Pairs Trading

A recipe for pairs trading is given here. However, its details may be different depending on the trading scenario and objectives of the trader.

1. Identify two stocks or a stock and an ETF within the same industry group with historically correlated price variations.
2. Using historical prices of the two stocks (or stock and ETF), $p_1(n)$ and $p_2(n)$, calculate their return processes, $r_1(n)$ and $r_2(n)$.
3. Estimate the beta, $\hat{\beta}$, and the residual process, $\hat{\epsilon}(n)$ of the model defined in (3.5.1) using the least-squares algorithm (3.5.4) (or some other estimator of choice) from actual market data.
4. Calculate the spread as the cumulative sum of the residual process, $\hat{\epsilon}(n)$, according to (4.5.9), expressed as

$$\Delta(n) \triangleq \sum_{j=1}^{W} \hat{\epsilon}(n - W + j),$$

   where W is the number of samples in the estimation window.
5. Calculate the mean and standard deviation of the spread, μ_Δ and σ_Δ, respectively. Then, normalize the spread as

$$\bar{\Delta}(n) = \frac{\Delta(n) - \mu_{\Delta}}{\sigma_{\Delta}}$$

6. If we do not have a position in the spread at time n, we have three options.
   (a) If $\bar{\Delta}(n) > T_1$, where T_1 is a predefined threshold (to open a position), then, we *short sell the spread* (open a *short* position in the spread, i.e., *short sell first stock* and *buy second stock*).
   (b) If $\bar{\Delta}(n) < -T_1$, then, we *buy the spread* (open a *long* position in the spread, i.e., *buy first stock* and *short sell second stock*).
   (c) Otherwise, we do nothing.

7. If we have a *short* or *long* position in the spread at time n and $|\bar{\Delta}(n)| < T_2 < T_1$ then we get out of the position in the spread. T_2 is another predefined threshold *to close an open position* in the spread. We get out of a short position in the spread (*buy first stock to cover* and *sell second stock*) when $\bar{\Delta}(n) < T_2$. Similarly, we get out of a long position in the spread (*sell first stock* and *buy second stock to cover*) when $\bar{\Delta}(n) > -T_2$.

As we discussed before, we do not re-estimate the parameters of the strategy during an open position and they are used until a trade is finished. See file `pairs_trading.m` for the MATLAB implementation of the recipe for pairs trading.

4.6 STATISTICAL ARBITRAGE

According to the Oxford Dictionary, *arbitrage* means "the simultaneous buying and selling of securities, currency, or commodities in different markets or in derivative forms in order to take advantage of differing prices for the same asset." In *statistical arbitrage* (also known as *StatArb*), we form the spread and make decisions to buy or sell it according to mathematical models, statistical measurements and analysis. Statistical arbitrage is also described as "highly technical short-term mean-reversion strategies involving large numbers of securities (hundreds to thousands, depending on the amount of risk capital), very short holding periods (measured in seconds to days), and substantial computational, trading, and information technology (IT) infrastructure" in [46]. In this section, by utilizing our knowledge of factor models (Section 3.5.3) and pairs trading (Section 4.5), we study a statistical arbitrage trading strategy where we look for potential arbitrage opportunities due to the price inefficiencies in the market tracked through the variations in the correlation structure among stocks in a portfolio (basket).

In Section 4.5, we developed a trading strategy based on the relative value model given in (3.5.1). Statistical arbitrage may be considered as a generalized version of the pairs trading where we use an extended model, like the factor model given in (3.5.5), to develop a trading strategy [29]. According to (3.5.5), the return of the jth stock is modeled as

$$r_j(n) = \beta_{j,1}f_1(n) + \beta_{j,2}f_2(n) + \ldots + \beta_{j,N}f_N(n) + \xi_j(n), \tag{4.6.1}$$

$$= \sum_{i=1}^{N} \beta_{j,i}f_i(n) + \xi_j(n) \tag{4.6.2}$$

where $f_i(n)$ is the ith factor at n and $\beta_{j,i}$ is the factor loading (weighting coefficient) in the return of the jth stock for the ith factor. We note that the right side of this equation is comprised of the prediction with respect to the factors as given in the summation and the prediction error (noise) term, $\xi_j(n)$. Then, similar to the pairs trading scenario, one invests 1 dollar (long) in stock j and $-\beta_{j,i}$ dollars (short) in factor i at n in this case, for all i and j values (assuming the factor is tradable). Therefore, ignoring the transaction costs and interest on cash for simplicity, return on investment for stock j is expressed as

$$r_{\text{inv},j}(n) = r_j(n) + (-\beta_{j,1})f_1(n) + (-\beta_{j,2})f_2(n) + \ldots + (-\beta_{j,N})f_N(n).$$

$$= r_j(n) - \sum_{i=1}^{N} \beta_{j,i}f_i(n) \tag{4.6.3}$$

Substitution of (4.6.1) into (4.6.3) yields

$$r_{\text{inv},j}(n) = \xi_j(n). \tag{4.6.4}$$

According to this equation, in this trading strategy, return on investment for a particular stock depends on only its excess return over a set of factors with their weights. As we describe in pairs trading (Section (4.5)), one only enters in such a trade when there is an *arbitrage* opportunity. It means the spread between the stock and the weighted factors (cumulative sum of the excess returns, $\xi_j(n)$) is large and expected to revert back to its mean value in the near future. Moreover, the factors, $f_i(n)$, are generated in a *statistical* sense. Hence, it is called *statistical arbitrage*. In the statistical arbitrage strategies, we hedge our trades with the properly weighted factors.

The factors we choose dictate the trading strategy. The requirements for the factors are listed as follows.

1. Factors must be generated statistically out of market data. We are not interested in deterministic factors such as return of a risk-free asset or fundamental factors like unemployment rate within the last year.
2. Factors must be tradable. We need to be able to trade actual financial instruments to realize (4.6.4) in order to take advantage of an arbitrage opportunity available in the market.
3. Factors should be in limited numbers in order to keep the transaction (trading) costs within a reasonable range.
4. Factors should be statistically independent or, at the very least, pairwise uncorrelated.

A widely used type of factors that satisfy these requirements are principal components (eigenvectors) of the empirical correlation matrix of stock returns in a portfolio [29]. Such factors are called *eigenportfolios* (3.5.12) as they are generated statistically, tradable, and in general, a small number of associated eigenvalues explains a significant variance part of the covariance structure of the stock returns in the portfolio. Furthermore, the returns of eigenportfolios are pairwise uncorrelated and such factors lead to effective regression models (3.5.13). However, use of eigenportfolios in statistical arbitrage strategies comes with two immediate problems that one needs to tackle with. Namely, they are transaction costs and risk management.

Trading eigenportfolios may be costly since the trade prescribed in (4.6.3) involves buying or selling of N stocks twice (once to get in and once to get out of the trade). Usually, N is a large number since one would pick a wide selection of stocks for the trading universe in order to be able to create eigenportfolios explaining the returns of a broader market. Therefore, the arbitrage may not cover the prohibitive transaction costs. Moreover, spread between any stock in the trading universe and the eigenportfolios may be large on either side with respect to its statistical equilibrium. This may stress the allowable risk budget of the trading strategy. From (4.6.3), we have the total return on investment expressed as

$$\begin{aligned} r_{\text{inv}}(n) &= \sum_{j=1}^{N} q_j(n) r_{\text{inv},j}(n) \\ &= \sum_{j=1}^{N} q_j(n) \left[r_j(n) - \sum_{i=1}^{N} \beta_{j,i} f_i(n) \right] \\ &= \sum_{j=1}^{N} q_j(n) r_j(n) - \sum_{i=1}^{N} \sum_{j=1}^{N} q_j(n) \beta_{j,i} f_i(n) \end{aligned} \tag{4.6.5}$$

where $q_j(n)$ is the investment in stock j at time n as defined in (4.4.5). In theory, realization of the investment described by (4.6.5) for a single trade of spread may actually require $2N$ trades [29].

In addition to transaction cost, one needs to measure and manage the investment risk in statistical arbitrage built on eigenportfolios as regression factors. Therefore, one aims to maximize the investment return for a predefined risk associated as

$$\sigma_{\text{inv}} = \left(E\left\{ r_{\text{inv}}^2(n) \right\} - E\left\{ r_{\text{inv}}(n) \right\}^2 \right)^{1/2}. \tag{4.6.6}$$

Substitution of (4.6.5) into (4.6.6) is easily traceable as we show it in Section 5.1.4 in detail. However, maintenance of a trade book while addressing all concerns raised here is not always trivial in real life.

4.6.1 A Recipe for Statistical Arbitrage

We provide a recipe for statistical arbitrage employing eigenportfolios as factors in the following.

1. Create a trading universe of N stocks based on some metrics of choice.
2. Calculate the return processes, $r_i(n)$, by using historical stock prices $p_i(n)$ $1 \le i \le N$.
3. Calculate the empirical correlation matrix, $\mathbf{P}$, defined in (3.2.12).
4. Calculate the eigenvalues, λ_k where $1 \le k \le N$ and $\lambda_k > \lambda_{k+1}$, and corresponding eigenvectors, ϕ_k, of $\mathbf{P}$ according to (3.5.9).
5. Estimate the volatilities of stocks in the trading universe, σ_i, according to (3.2.4).
6. Calculate the eigenportfolio returns $f_k(n)$ in (3.5.12) by using ϕ_k, σ_i, and $r_i(n)$ for the largest L eigenvalues, $1 \le k \le L \le N$.
7. Estimate the beta, β, and the residual process, $\xi(n)$, of the model defined in (4.6.1) by using (3.5.14) (or another estimator of choice).
8. The rest of the recipe is identical to the pairs trading. Hence, just follow the steps starting with Step 4 in Section 4.5.3 to generate signals for potential trades.

We remind that we are not entertaining a new trading signal to open a position if we already have identical existing position in the same stock. See file `statistical_arbitrage.m` for the MATLAB implementation of the recipe for statistical arbitrage explained in this section. See [29] for further details on statistical arbitrage in trading stocks.

4.7 TREND FOLLOWING

In stock markets, one often observes an up or down price *trend* (or *momentum*) of a stock at a given time scale like every day or hour (also called trading frequency or rebalancing period). It is quite common for the price of a stock to gain or lose its market value consistently for a prolonged time. Such a trend corresponds to the fat-tails in price process $p(n)$ as well as in drift parameter, μ, of the geometric Brownian motion model described in (3.1.2). In this strategy, we *follow* the trend (momentum) and jump on it by buying and selling the stock at the right time to generate profit from the trade. If our analysis signals that the price is at the beginning of an up trend (or at least the price is exhibiting a strong upwards momentum) we buy the stock. When the up trend ends (or at least the upward momentum weakens) we exit the position by selling the stock. For the down trend, the start the trade by short selling the stock and finishing it by buying it back to cover for the short position.

In this section, for the analysis step of the trend following strategy, we present and discuss the *moving averages* that are the most commonly used trend *indicators*. We note that trend can be detected through various other techniques, e.g., wavelets [47], neural networks [48, 49], hidden Markov models [50], and others. Then, we study methods that utilize the moving averages to generate trading signals. Finally, we discuss moving averages through the perspective of linear shift-invariant (LSI) discrete-time filters that facilitates us to study their frequency response and to design more sophisticated filters.

We note that trend following strategies are usually questioned by the quantitative finance community. It is mostly due to the fact that these strategies are alluded to technical trading or *charting* in which traders draw geometric shapes (trend lines, support, resistance, etc.) on the price charts. Decisions made based on those shapes are naturally highly subjective. However, *trend* and *seasonality* in general are well established concepts in the time series analysis literature. Some people argue that technical indicators, when executed systematically, provide incremental information and offer some practical value [51]. According to a survey on profitability of technical indicators [52], 56 studies among the total of 95 show profits with technical trading strategies. Furthermore, 20 of them generate negative results, and 19 studies indicate mixed performance results.

4.7.1 Moving Averages

Traders use averaging to filter out the "noise" in price data that allows them to better focus on the trend. There are variations of *moving averages* (MA). The simplest one is the *simple moving average* (SMA) in which the average of the last N price samples at n is calculated as follows:

$$\mathrm{SMA}(n) = \frac{1}{N}\sum_{k=0}^{N-1} p(n-k), \tag{4.7.1}$$

where $p(n)$ is the price at time n. SMAs (MAs in general) are named based on the sampling period and the window size N. For example, if the sampling (rebalancing) period is a day, an MA with $N=20$ is called as "20 day moving average." A modified version of SMA is the *exponential moving average* (EMA) where the most recent price observations have the highest weights in the average calculation. EMA is defined as

$$\mathrm{EMA}(n) = \alpha p(n) + (1-\alpha)\,\mathrm{EMA}(n-1), \tag{4.7.2}$$

with $\mathrm{EMA}(0) = p(0)$ where α is *the memory parameter* with $0 < \alpha < 1$. Therefore, the higher the α the quicker we forget about the old price data. The daily price for the Apple Inc. (AAPL) stock between September 1, 2013 and August 31, 2014, its SMA with $N = 20$ and $N = 50$, and its EMA with $\alpha = 0.095$ are displayed in Figure 4.7.1. See file `moving_average.m` for the implementation of SMA and EMA in MATLAB.

4.7.2 Signal Generation Methods for Trend Following

The simplest method to generate signals for trend following strategy is called *crossover*, in which we long the stock (or exit the open short position then long the stock) if the price of the stock, $p(n)$, crosses above the moving

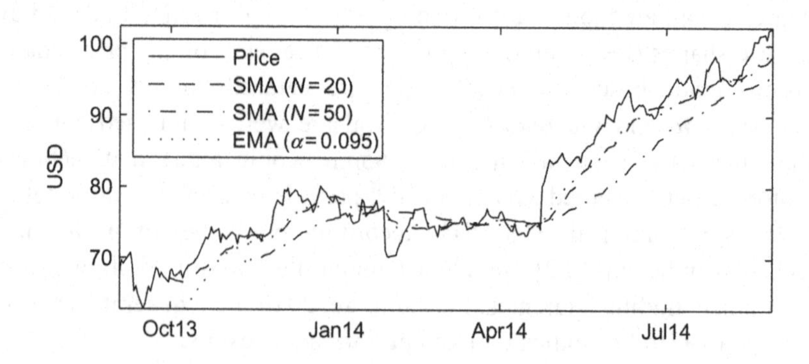

Figure 4.7.1 Daily price for the Apple Inc. (AAPL) stock between September 1, 2013 and August 31, 2014 as well as the corresponding SMA for N = 20 and N = 50 and EMA for α = 0.095.

average (MA). In other words, if price was below MA and now it is above MA, $p(n) > \text{MA}(n)$ and $p(n-1) < \text{MA}(n-1)$, it indicates an upward trend and we long the stock. Similarly, if the price crosses below MA, $p(n) < \text{MA}(n)$ and $p(n-1) > \text{MA}(n-1)$, then, we short sell the stock (or exit the long position in the stock then short sell it).

However, small deviations in the price, related to its volatility σ in (3.1.2), may create high levels of measurement noise and we get in and out of positions too quickly. This causes significant increase in cost due to over-trading (Section 4.3). Therefore, a modified strategy is called *crossover with a filter*. The word "filter" in this context refers to any measure that validates the trading signal. Some scenarios are described as follow.

1. It is an indication of an up (down) trend when *fast moving average* of the price, FMA with a small time window $N = N_{\text{f}}$ in (4.7.1), crosses above (below) the *slow moving average* of the price, SMA with a large time window $N = N_s$ in (4.7.1) where $N_s > N_{\text{f}}$.
2. It is an indication of an up (down) trend when price crosses above (below) the 5% (or some other percentage) of its moving average.
3. It is an indication of an up (down) trend when *fast moving average* of the price crosses above (below) $k\sigma_{\text{p}}(n)$ plus (minus) SMA where k is a constant and $\sigma_{\text{p}}(n)$ is the empirical standard deviation of the price (usually estimated with the same window used for the *slow moving average*).

There is a commonly used technical trading indicator named *Bollinger Band®* that utilizes exponential moving average defined in (4.7.2). However, in contrast to the ones discussed above, it is a *contra-trend* indicator. The bands are defined as $\text{EMA}(n) \pm k\sigma_{\text{p}}(n)$. When price touches the upper (lower) band, it is an indication that the stock is overbought (oversold) and a short-term down (up) trend is expected.

4.7.3 Moving Averages as Discrete-Time Filters

Let us represent *simple moving average* as the output of a linear shift-invariant (LSI) discrete-time filter as

$$\begin{aligned}\text{SMA}(n) &= h(n) * p(n)\\ &= \sum_{m=-\infty}^{\infty} h(m)p(n-m),\end{aligned}$$

where $*$ is the linear convolution operator and $h(n)$ is the *unit sample response* of the LSI filter as given

$$h(n) = \begin{cases} \frac{1}{N} & 0 \le n \le N-1 \\ 0 & \text{otherwise} \end{cases}. \tag{4.7.3}$$

The frequency response of the filter is the discrete-time Fourier transform (DTFT) of $h(n)$ (4.7.3) calculated as

$$\begin{aligned} H\left(e^{j\omega}\right) &= \mathcal{F}\{h(n)\} \\ &= \sum_{n=-\infty}^{\infty} h(n)e^{-j\omega n}, \end{aligned} \tag{4.7.4}$$

where j is the imaginary unit, $\omega = 2\pi f$ is the angular frequency, and f is the frequency in Hertz. Substitution of (4.7.3) into (4.7.4) and following through the algebraic expressions the frequency response of the moving average filter is found as [53]

$$H\left(e^{j\omega}\right) = e^{-j\omega(N-1)/2}\frac{1}{N}\frac{\sin(N\omega/2)}{\sin(\omega/2)}. \tag{4.7.5}$$

It is a periodic function in ω with a period of 2π. The magnitude response of the moving average filter is expressed as

$$\left|H\left(e^{j\omega}\right)\right| = \frac{1}{N}\frac{\sin(N\omega/2)}{\sin(\omega/2)}.$$

The magnitude function of the frequency response has zero crossings at $2\pi k/N$ with $k = 1, 2, \ldots, \lfloor N/2 \rfloor$ in a period (4.7.5). These zero crossings result in a 180° jump in phase response. Therefore, one can express the phase response of the moving average filter as

$$\theta(\omega) = \angle H\left(e^{j\omega}\right) = -\frac{N-1}{2}\omega + \pi \sum_{k=1}^{\lfloor N/2 \rfloor} u\left(\omega - \frac{2\pi k}{N}\right),$$

where $\lfloor \cdot \rfloor$ is the floor operator and $u(\omega)$ is the step function in ω. Magnitude and phase responses of the moving average filters for $N = 5$ and $N = 20$ are displayed in Figure 4.7.2. The group delay of the moving average filter is given as [53]

$$\tau(\omega) = -\frac{d\theta(\omega)}{d\omega} = \frac{N-1}{2},$$

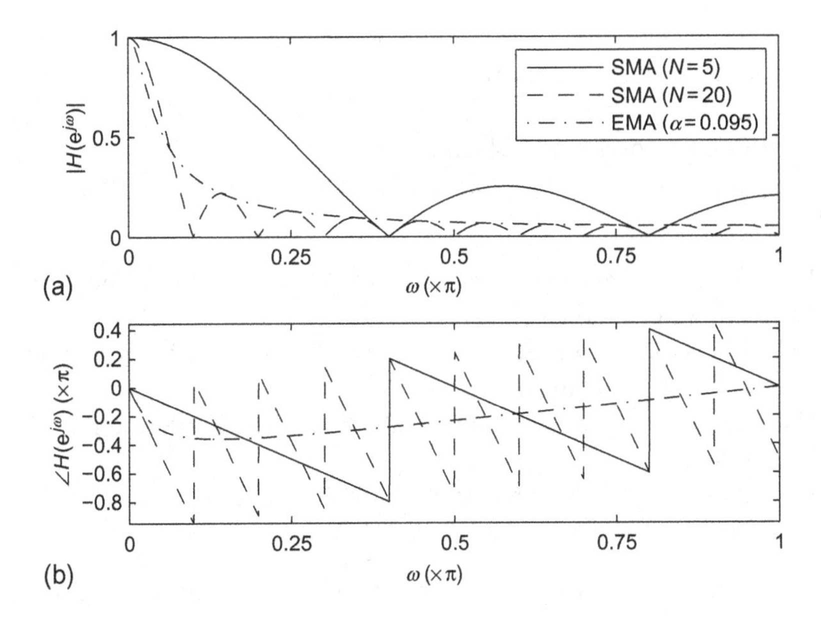

Figure 4.7.2 (a) Magnitude and (b) phase responses of SMA filter for $N = 5$, $N = 20$ *and EMA filter with* $\alpha = 0.095$ *and* $N = 20$.

that is a constant. It is observed that SMA filter has a constant group delay of approximately half the filter length. Therefore, effect of a change in trend can only be detected after $(N - 1)/2$ samples (trading periods).

Similarly, we can represent exponential moving average (EMA) defined in (4.7.2) as an LSI filter with the output

$$\mathrm{EMA}(n) = h(n) * p(n),$$

and its transfer function

$$H(\mathrm{e}^{j\omega}) = \mathcal{F}\{h(n)\} = \frac{\alpha}{1 - (1 - \alpha)\,\mathrm{e}^{-jw}}.$$

The magnitude and phase responses of an EMA filter with $\alpha = 0.095$ and $N = 20$ are displayed and compared with SMA filter in Figure 4.7.2. It is observed from the figure that EMA does not have a constant group delay since its phase response is non-linear.

Investigation of moving averages for trading application from the LSI filter perspective enables us to utilize filter design techniques available in the digital signal processing literature. The nature of the price signal and what we look for dictate the design specs. For example, the design of a variation of the exponential moving average filter, named as "the instantaneous trend-line," with zero group delay at $\omega = 0$ and low group delay for small values of ω is given in [54].

4.7.4 A Recipe for Trend Following

A straightforward recipe for trend following strategy is summarized as follows.

1. Identify the stock and the sampling (trading) frequency that exhibit strong up and down price trends.
2. Calculate its moving average by using (4.7.1) or (4.7.2) (or any other method to determine trends).
3. Use one of the signal generation methods discussed in Section 4.7.2 to identify trends.
4. If it is an up (down) trend, then, buy (short sell) the stock. If trend weakens or starts to go in the opposite direction, exit the position.

See file `trend_following.m` for the MATLAB implementation of the recipe for trend following strategy.

4.8 TRADING IN MULTIPLE FREQUENCIES

The data we use in the analysis phase is sampled at a certain sampling (trading) frequency. We usually use the same time interval (frequency) for data sampling and for trading (rebalancing). For example in a pairs trading strategy, we acquire the daily price data (sampling interval/frequency is 1 day) for two stocks, and then calculate their daily spread. Then, if the spread is over or under a threshold for that day, we decide to get in or get out of a position in that spread. We probably trade more than one pair. Although we trade different pairs, not necessarily on the same day, we trade *all* of them at the *same frequency*. However, it is argued that one may also seek to trade *multiple* stocks in a portfolio in *multiple frequencies* [41]. This trading perspective involves the use of market data sampled at different frequencies for different stocks as well. The intuition to trade in multiple frequencies is highlighted as follows.

1. Liquidity at any given time, the availability of the stock in the market, may not be the same for all stocks of the portfolio. Therefore, the trader may want to trade certain stocks faster or slower than the others.
2. Different stocks may reveal certain aspects of the market such as a trend or a relative-value at different sampling frequencies.

Although some strategies like trend following do not need to use correlation structure, risk management methods often utilize it as discussed in Chapter 5. Therefore, we need to measure and monitor variations of pairwise return correlations with the proper adjustments due to varying holding times for different stocks when trading in multiple frequencies.

Assuming that price follows a geometric Brownian motion model defined in (3.1.2), it follows from our discussions in Section 3.2.2 that volatilities estimated at different sampling frequencies (holding times or horizons) have the following relationship

$$\sigma_1 = \sqrt{k_1/k_2}\sigma_2 = m\sigma_2, \tag{4.8.1}$$

where σ_1 and σ_2 are the volatilities estimated at sampling intervals $k_1 T_s$ and $k_2 T_s$, respectively, and T_s is the baseline sampling interval. Therefore, any trading scenario in which different frequencies are used employs the constant m given in (4.8.1) to adjust for the common sampling period (holding time). We discuss in detail the risk management for trading in multiple frequencies in Section 5.2.

4.9 SUMMARY

Traders seek profits in the short term by identifying inefficiencies and strong trends in asset prices and speculating on their future performance based on analysis of currently available historical data. They build trading strategies that analyze market data, create signals based on a set of rules to open or close a long or short position in a particular asset in an investment portfolio. These strategies are tested using backtesting procedures before they go live trading in markets. A good backtesting framework involves models for market impact and trading cost along with investment risk. Backtesting simulates and measures the performance of a strategy for historical market data. The most commonly used metric to measure the performance of a trading strategy is the Sharpe ratio, the risk adjusted return of the portfolio equity, identified by the profit and loss (P&L) equation. Trading strategies may be grouped into two major categories. They are called mean-reversion and trend-following strategies. The pairs trading that utilizes the spread between two stocks (asset returns) to trade is the first mean-reversion strategy presented in the chapter. The second mean-reversion strategy covered here is a statistical arbitrage where eigenportfolios are used in order to identify over and undervalued stocks in a trading universe. There are many trend following strategies that seek upward and downward momentum in the price of a stock. We presented the most trivial trend following strategy where the crossovers of the fast and slow moving averages of the asset price generate trading signals. A good trader has a blend of skill set including a thorough understanding of market macro and microstructures with years of experience quantified by record of successful trading performance.

CHAPTER 5

Risk Estimation and Management

An investment portfolio is comprised of various assets with their own specific risk and return characteristics. The diversity of a portfolio facilitates the portfolio manager with a level of freedom to maintain it according to a preset target performance. Therefore, designing an optimal portfolio for the desired specification and its maintenance (rebalancing) in time require financial acumen along with expertise in econometrics and risk management. In order to estimate the risk of a portfolio with N assets defined in (3.3.7), we need to estimate $N(N-1)/2$ cross-correlations of pairwise asset returns to form its $N \times N$ empirical correlation matrix, $\mathbf{P}$ (3.2.12). It is known that, $\hat{\mathbf{P}}$ contains a significant amount of inherent measurement noise due to market microstructure that needs to be removed. Eigenfiltering has been successfully employed to filter out this undesirable noise component from the measured correlations [41, 55–57]. In this chapter, we discuss in detail the eigenfiltering of $\hat{\mathbf{P}}$ for better risk estimation of a portfolio. We also introduce approximations to the correlation matrix for efficient noise filtering. Then, we revisit a straightforward risk management method and its two modifications. We conclude the chapter by demonstrating the performance improvements due to the modifications in the original method.

5.1 EIGENFILTERING OF NOISE IN EMPIRICAL CORRELATION MATRIX

In this section, we focus on the intrinsic measurement noise in the empirical correlation matrix of asset returns in a portfolio and its removal by eigenfiltering, using eigenanalysis of the matrix to filter out the noise component

A Primer for Financial Engineering. http://dx.doi.org/10.1016/B978-0-12-801561-2.00005-8

for a more robust risk estimation. We start our discussion by introducing the asymptotic distribution of eigenvalues for a random matrix that gives us a mathematical framework to handle such a problem.

5.1.1 Asymptotic Eigenvalue Distribution of a Random Matrix

Let us construct a random matrix $\mathbf{K}$ of size $N \times N$ as

$$\mathbf{K} = \frac{1}{M}\mathbf{W}^{\mathrm{T}}\mathbf{W}, \tag{5.1.1}$$

where $\mathbf{W}$ is an $M \times N$ matrix comprised of uncorrelated elements drawn from a Gaussian distribution with zero mean and variance σ^2, $[W_{m,n}] \sim \mathcal{N}(0, \sigma^2)$, where $E\{W_{m,n}W_{m+k,n+l}\} = \sigma^2\delta_{m-k}\delta_{n-l}$ for $m = 1, 2, \ldots, M$ and $n = 1, 2, \ldots, N$. We note that $\mathbf{K}$ belongs to the family of Wishart matrices as referred in the multivariate statistical theory. Statistics of random matrices such as $\mathbf{K}$ are extensively studied in the literature [57]. It is known that the distribution of the eigenvalues of the random matrix $\mathbf{K}$ in the limit is expressed as [58]

$$f(\lambda) = \frac{M}{2\pi\sigma^2 N}\frac{\sqrt{(\lambda_{\max} - \lambda)(\lambda - \lambda_{\min})}}{\lambda}, \tag{5.1.2}$$

where $f(\cdot)$ is the *probability density function* (pdf), $N \to \infty$, $M \to \infty$ with the ratio M/N fixed, and $\lambda_{\max}$ and $\lambda_{\min}$ are the maximum and minimum eigenvalues of $\mathbf{K}$, respectively, for the same limit case, with the values defined as [58]

$$\lambda_{\max,\min} = \sigma^2\left(1 + \frac{N}{M} \pm 2\sqrt{\frac{N}{M}}\right). \tag{5.1.3}$$

Therefore, for a random matrix, constructed according to (5.1.1), we know the distribution of its eigenvalues in closed-form as given in (5.1.2). We utilize this expression during the discussion of the built-in noise in the empirical correlation matrix of asset returns as presented in the next section.

5.1.2 Noise in the Empirical Correlation Matrix

We estimate the empirical correlation matrix $\mathbf{P}$ of (3.2.12) as

$$\hat{\mathbf{P}} = \frac{1}{M}\bar{\mathbf{R}}^{\mathrm{T}}\bar{\mathbf{R}}, \tag{5.1.4}$$

where $\bar{\mathbf{R}}$ is the $M \times N$ historical asset return matrix, N is the number of assets in the portfolio, and M is the number of available return samples per asset. Each element of $\bar{\mathbf{R}}$, $\bar{R}_{mn}$, is the normalized return of the nth asset at the mth discrete time instance. We do the normalization such that the return time series of each asset, each column of $\bar{\mathbf{R}}$, is zero mean and unit variance. The element located on the ith row and jth column of $\hat{\mathbf{P}}$ is equal to

$$\hat{P}_{ij} = \frac{1}{M}\sum_{m=0}^{M-1} \bar{r}_i(m)\bar{r}_j(m), \tag{5.1.5}$$

where $\bar{r}_i(m) = \bar{R}_{mi}$ is the normalized return of the ith asset at the mth discrete time instance as defined

$$\bar{r}_i(m) = \bar{R}_{mi} = \frac{r_i(m) - \hat{\mu}_i}{\hat{\sigma}_i}, \tag{5.1.6}$$

where $r_i(m)$ is the return of the ith asset at the mth discrete time, $\hat{\mu}_i$ is the estimated mean of $r_i(m)$ as defined

$$\hat{\mu}_i = \frac{1}{M}\sum_{m=0}^{M-1} r_i(m), \tag{5.1.7}$$

and $\hat{\sigma}_i$ is the estimated standard deviation of $r_i(m)$ shown as

$$\hat{\sigma}_i = \left(\frac{1}{M-1}\sum_{m=0}^{M-1}\left[r_i(m) - \hat{\mu}_i\right]^2\right)^{\frac{1}{2}}. \tag{5.1.8}$$

There are several points to make here. First, if the return processes were stationary, choosing M as large as possible would be the best approach to improve the estimation. However, in financial processes, anything is hardly stationary. Second, some assets may not have long history. For example, centuries-long historical data for cotton price might be available whereas Internet-based assets have been around for only about two decades. Third, we might want to exploit the short-term impacts of certain economic crises periods in time. However, choosing a very long estimation time window would wipe out time-local events that are significant. For these reasons, choosing the time window, M, is not a trivial task and deserves special attention for tuning of mathematical models. We provide further insights for the parameter selection in the following example.

Example 5.1. We know that if $\hat{\mathbf{P}}$ were a random matrix, its eigenvalues would be samples drawn from the distribution given in (5.1.2). We choose the sampling period of 15 min for the price, and use a time window between January 4, 2010 and May 18, 2010. Hence, $M = 2444$ (since we have 94 business days and 26 data samples per day in that interval). Assets in the portfolio are the 494 of 500 stocks listed in S&P 500 index as of January 4, 2010. Thus, we have $N = 494$ and $M/N = 4.95$. Next, we decompose

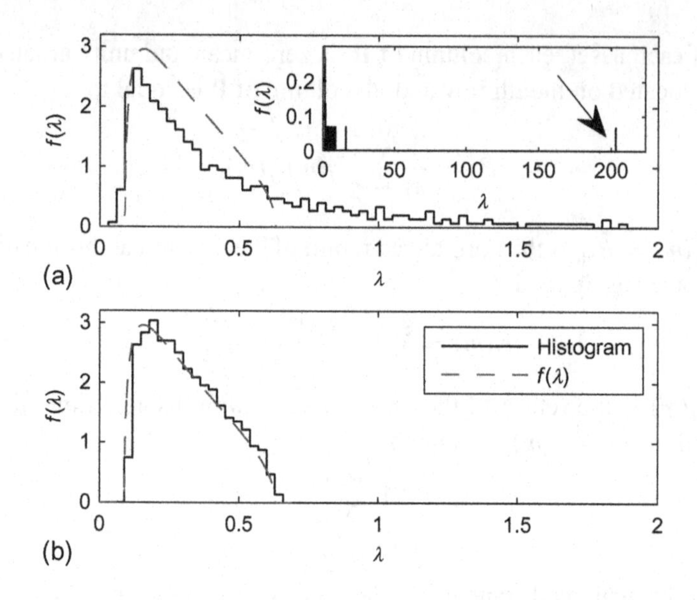

Figure 5.1.1 (a) Histogram of the eigenvalues for the empirical correlation matrix along with the pdf in the limit of the eigenvalues for a random matrix (5.1.2). *(b) Histogram of the eigenvalues for an empirical random matrix* (5.1.1) *along with the pdf in the limit.*

matrix $\hat{\mathbf{P}}$ into its eigenvectors with their corresponding eigenvalues. We calculate the histogram of the eigenvalues [55, 56] and display them in Figure 5.1.1a along with the probability density function of the eigenvalues of a random matrix expressed in (5.1.2), and calculated with $M/N = 4.95$, $\sigma^2 = 0.3$, $\mu = 0$, $\lambda_{\max} = 0.63$, and $\lambda_{\min} = 0.091$ according to (5.1.3). We infer from the figure, by setting the parameter $\sigma^2 = 0.3$ in (5.1.2), that a reasonable fit to the empirical data for eigenvalues smaller than 0.63 is achievable. This suggests that about 30% of the energy in matrix $\hat{\mathbf{P}}$ is random. Hence, for this example, we can consider the eigenvalues smaller than 0.091 as noise. We also note that, for this example, the largest 60 of the 494 eigenvalues, 13%, represent about 70% of the total variance in $\hat{\mathbf{P}}$. Moreover, only the largest four eigenvalues represent 50% of the total energy. The maximum and minimum eigenvalues, $\lambda^{\mathrm{P}}_{\max}$ and $\lambda^{\mathrm{P}}_{\min}$, are equal to 203.24 and 0.044, respectively. The largest eigenvalue is approximately 322 times larger than its counterpart in a random matrix. We see from Figure 5.1.1a. that histogram of eigenvalues for $\hat{\mathbf{P}}$ has two major clusters. Namely, there is a bulk of eigenvalues that must be strongly related to the noise, and the remaining relatively small number of eigenvalues deviating from the bulk that must be representing the valuable information. Since 95% of the eigenvalues have their values less than 2, we zoom into the region bounded by $0 \leq \lambda \leq 2$

and display the full histogram on the top right corner in Figure 5.1.1a. We mark the largest eigenvalue, 203.24, with an arrow. For validation purposes, we generate a random matrix **K** of (5.1.1) with the parameters $M/N = 4.95$, $\sigma^2 = 0.3$, and $\mu = 0$. We display the histogram of the eigenvalues for **K** along with the probability density function of the eigenvalues for a random matrix as defined in (5.1.2) using the same parameters in Figure 5.1.1b. We observe from the figure that the empirical histogram fits quite nicely to the theoretical limiting distribution as described in this section.

5.1.3 Eigenfiltering of Built-in Market Noise

The market microstructure and price formation process of an asset naturally introduces peculiar built-in noise in the measured market data. Let us decompose the empirical correlation matrix of asset returns (5.1.4) into its eigenvalues and eigenvectors as follows [27]

$$\hat{\mathbf{P}} = \mathbf{\Phi}\mathbf{\Lambda}\mathbf{\Phi}^{\mathrm{T}}, \tag{5.1.9}$$

where $\mathbf{\Lambda} = \mathrm{diag}(\lambda_k)$ is a diagonal matrix and λ_k is the kth eigenvalue in the descending order $\lambda_k \geq \lambda_{k+1}$, and $\mathbf{\Phi}$ is an $N \times N$ matrix comprised of N eigenvectors as its columns as given

$$\mathbf{\Phi} = \begin{bmatrix} \boldsymbol{\phi}_1 & \boldsymbol{\phi}_2 & \cdots & \boldsymbol{\phi}_N \end{bmatrix}, \tag{5.1.10}$$

and $\boldsymbol{\phi}_k$ is the $N \times 1$ eigenvector corresponding to the kth eigenvalue λ_k with the intrinsic properties $\lambda_k \geq 0\ \forall_k$, $\sum_k \lambda_k = N$. Moreover, $\mathbf{\Phi}\mathbf{\Phi}^{\mathrm{T}} = \mathbf{I}$ due to the orthonormality property of the eigenvectors. By substituting (5.1.9) into (3.3.7), we formulate the estimated risk of the portfolio as [41]

$$\hat{\sigma}_{\mathrm{p}} = \left(\mathbf{q}^{\mathrm{T}}\mathbf{\Sigma}^{\mathrm{T}}\mathbf{\Phi}\mathbf{\Lambda}\mathbf{\Phi}^{\mathrm{T}}\mathbf{\Sigma}\mathbf{q}\right)^{1/2}. \tag{5.1.11}$$

Following the approach proposed in [56], we define the *eigenfiltered correlation matrix* as

$$\tilde{\mathbf{P}} = \sum_{k=1}^{L} \lambda_k \boldsymbol{\phi}_k \boldsymbol{\phi}_k^{\mathrm{T}} + \mathbf{E}, \tag{5.1.12}$$

where L is the number of selected eigenvalues satisfying $\lambda_k \geq \lambda_{\max}$ (5.1.3), and $L \ll N$. We note that, identifying L is not a trivial task. Although (5.1.3) offers a framework to calculate it and some insights provided in Example 5.1, in practice, professional investors heavily backtest (Section 4.4) to fine tune values of such parameters, like L here, used in financial models and frameworks in order to achieve the best possible performance. We highlight that we introduced the diagonal matrix **E** in

(5.1.12) to preserve the total variance, helping to keep the traces of $\tilde{\mathbf{P}}$ (5.1.12) and $\hat{\mathbf{P}}$ (5.1.9) equal. We define matrix $\mathbf{E}$ as

$$\mathbf{E} = \left[E_{ij}\right] = \varepsilon_{ij} = \begin{cases} 1 - \sum_{k=1}^{L} \lambda_k \phi_i^{(k)} \phi_j^{(k)} & i = j, \\ 0 & i \neq j, \end{cases} \tag{5.1.13}$$

where $\phi_i^{(k)}$ is the ith element of the kth eigenvector. We note that adding the matrix $\mathbf{E}$ in (5.1.12) is equivalent to setting the diagonal elements to be 1, $\left[\tilde{P}_{ii}\right] = 1$, $i = 1, 2, ..., N$. Now we can rewrite the filtered version of empirical correlation matrix from (5.1.12) and (5.1.13) as follows

$$\tilde{\mathbf{P}} = \left[\tilde{P}_{ij}\right] = \tilde{\rho}_{ij} = \sum_{k=1}^{L} \lambda_k \phi_i^{(k)} \phi_j^{(k)} + \varepsilon_{ij}. \tag{5.1.14}$$

By substituting the filtered version of empirical correlation, $\tilde{P}_{ij}$ in (5.1.14), into (3.3.7), we obtain the estimated risk as

$$\tilde{\sigma}_{\mathrm{p}} = \left[\sum_{k=1}^{L} \lambda_k \left(\sum_{i=1}^{N} q_i \phi_i^{(k)} \sigma_i \right)^2 + \sum_{i=1}^{N} \varepsilon_{ii} q_i^2 \sigma_i^2 \right]^{1/2}. \tag{5.1.15}$$

Following the similar steps provided in [55, 56], we experiment the impact of using noise filtered estimated risk (5.1.15) instead of noisy estimated risk (3.3.7) in the next example in order to better understand the concepts covered in the section. See file `risk.m` for the MATLAB code for the risk calculator given in (5.1.15).

Example 5.2. We have a data set comprised of return time series for 494 stocks listed in S&P 500 index. And we consider two time periods in the experiment. Namely, they are made of trading days from January 4, 2010 to May 18, 2010, and from May 19, 2010 to September 30, 2010. We assume that in the morning of May 19, 2010, the first day of the second time period, a risk manager has the perfect prediction of the future returns; the expected return vector of the second period, $\boldsymbol{\mu}_2$ of (3.3.6), is known. On that morning, the risk manager is asked to create a portfolio for a given target portfolio return, μ. First, the risk manager calculates the sample correlation matrix for the first time period, $\hat{\mathbf{P}}_1$, using (5.1.4). Then, the manager obtains the optimum investment allocation vector, $\mathbf{q}^*$, using (3.3.14) with $\hat{\mathbf{P}}_1$, $\boldsymbol{\mu}_2$, and μ. Next, the manager calculates the estimation of the predicted risk of the portfolio for the second time period, $\hat{\sigma}_2^{\mathrm{p}}$, using (3.3.7) with $\mathbf{q}^*$ and $\hat{\mathbf{P}}_1$. Finally, on September 30, 2010, the last day of the second time period, the manager obtains the empirical correlation matrix for the second time period, $\hat{\mathbf{P}}_2$ and estimation of the realized risk for the portfolio therein, $\hat{\sigma}_2^{\mathrm{r}}$, using (3.3.7) with $\mathbf{q}^*$ and $\hat{\mathbf{P}}_2$. Obviously, in this setup, the investment vector calculated through (3.3.14) is a function of the target portfolio return, μ, and the covariance matrix, $\mathbf{C} = \mathbf{\Sigma}^{\mathrm{T}}\mathbf{P}\mathbf{\Sigma}$. Moreover, portfolio risk calculated with (3.3.7) is a function of the investment allocation vector, $\mathbf{q}^*$, and the covariance matrix, $\mathbf{C}$. Hence, estimated portfolio risk is a function of the target portfolio return, μ, and the correlation matrix, $\mathbf{P}$. We formalize it as follows

$$\begin{aligned} \hat{\sigma}_2^{\mathrm{p}}(\mu) &= f\left(\hat{\mathbf{P}}_1, \mu\right), \\ \hat{\sigma}_2^{\mathrm{r}}(\mu) &= f\left(\hat{\mathbf{P}}_2, \mu\right), \end{aligned} \tag{5.1.16}$$

where $f(\cdot)$ defines a function that involves the calculation of the optimum investment vector defined in (3.3.14), and using it in (3.3.7) in order to obtain the portfolio risk. We display the plots of $\hat{\sigma}_2^{\mathrm{p}}(\mu)$ and $\hat{\sigma}_2^{\mathrm{r}}(\mu)$ as a function of μ in Figure 5.1.2a with solid and dashed lines, respectively. The risk manager also checks how large the risk was under- or over-estimated by defining an error function given as

$$\epsilon(\mu) = \left(\frac{\hat{\sigma}_2^{\mathrm{p}}(\mu)}{\hat{\sigma}_2^{\mathrm{r}}(\mu)} - 1\right) \times 100\%. \tag{5.1.17}$$

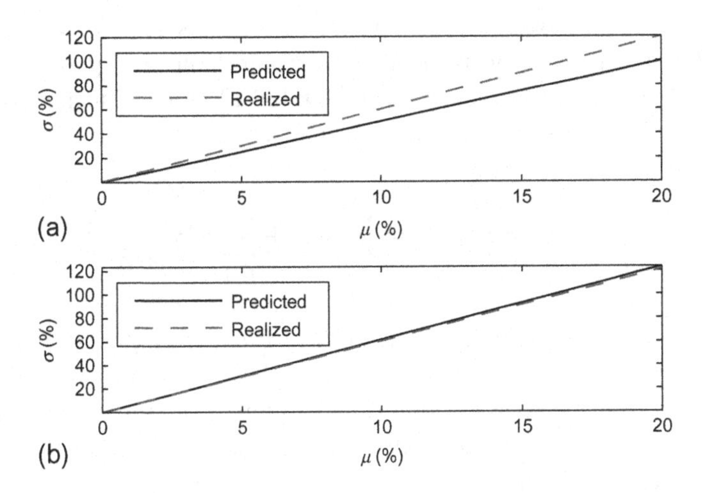

Figure 5.1.2 Predicted and realized risk functions versus target portfolio returns for the set of efficient portfolios where noisy empirical correlation matrices are used for the cases (a) without any filtering, and (b) with filtering prior to risk calculations.

Root mean square (RMS) value of this error function is ~16.8% for the case displayed in Figure 5.1.2a. In the second part of the example, risk manager does everything the same, but this time uses the filtered empirical correlation matrices $\tilde{\mathbf{P}}_1$ and $\tilde{\mathbf{P}}_2$, calculated by (5.1.12) with $L = 4$, prior to creating the portfolio and calculating the corresponding risk functions $\hat{\sigma}_2^{\mathrm{p}}(\mu)$ and $\hat{\sigma}_2^{\mathrm{r}}(\mu)$. We display the results for this case in Figure 5.1.2b. The RMS value of the error for this case is ~2.3%. We observe from the experiment that noise-free correlation matrix is more stable and it lets the risk manager predict the portfolio risk better than the noisy one.

Although the above example compares only two time periods and it is not 100% realistic due to the assumption that the future expected return vector $\boldsymbol{\mu}_2$ is known in advance by the risk manager, it emphasizes the significance of built-in noise in the correlation matrix and the importance of noise filtering to sanitize measured market data. In practice, the parameters of the system, such as the number of eigenvalues, L, need to be backtested over several time periods to build a high level of confidence in the parameter value for the given data of interest.

5.1.4 Estimation of Portfolio Risk in Statistical Arbitrage and Eigenfiltering of Market Noise

In most statistical arbitrage strategies (Section 4.6), each asset is associated with a number of assets in the investment portfolio based on certain metrics. Estimating the risk and eigenfiltering of the noise in the empirical correlation matrix for such a portfolio is not that trivial. In this section, we simplify the problem where every asset is traded against one *hedging asset*. For example, the hedging asset may be an ETF [29]. Return of such a portfolio is comprised of N assets and H hedging assets as given

$$r_{\mathrm{p}} = r_{\mathrm{a}} + r_{\mathrm{h}} = \sum_{i=1}^{N} q_i r_i + \sum_{j=1}^{H} g_j y_j, \tag{5.1.18}$$

where r_{a} and r_{h} are the total returns on N *investment assets* and H *hedging assets* (N and H may or may not be equal, in general), respectively; r_i and y_j are the returns of the ith asset and the jth hedging asset, respectively; and q_i and g_j are the amounts of capital invested in the ith asset and the jth hedging asset, respectively. In the rest of the section, for the clarity of discussion, we assume that the returns of the assets have zero mean noting that extension to non-zero mean case is trivial. We express the risk of the hedged portfolio as

$$\sigma_{\mathrm{p}} = E\left\{r_{\mathrm{p}}^2\right\}^{1/2}$$

$$= \left(\sigma_{\mathrm{a}}^2 + 2\sigma_{\mathrm{a}}\sigma_{\mathrm{h}} + \sigma_{\mathrm{h}}^2\right)^{1/2}$$

$$= \left(E\left\{r_{\mathrm{a}}^2\right\} + 2E\left\{r_{\mathrm{a}}r_{\mathrm{h}}\right\} + E\left\{r_{\mathrm{h}}^2\right\}\right)^{1/2}. \qquad (5.1.19)$$

It follows from (5.1.18) and (5.1.19) that we need to estimate:

1. Variance of total returns of *investment assets*, first term in (5.1.19), as given

$$\sigma_{\mathrm{a}}^2 = E\left\{r_{\mathrm{a}}^2\right\} = \sum_{i=1}^{N}\sum_{j=1}^{N} q_i q_j E\left\{r_i r_j\right\}. \qquad (5.1.20)$$

2. Cross-correlations between total returns of *investment assets* and total returns of *hedging assets*, second term in (5.1.19), as given

$$\sigma_{\mathrm{a}}\sigma_{\mathrm{h}} = E\left\{r_{\mathrm{a}}r_{\mathrm{h}}\right\} = \sum_{i=1}^{N}\sum_{j=1}^{H} q_i g_j E\left\{r_i y_j\right\}. \qquad (5.1.21)$$

3. Variance of total returns of *hedging assets*, third term in (5.1.19), as given

$$\sigma_{\mathrm{h}}^2 = E\left\{r_{\mathrm{h}}^2\right\} = \sum_{i=1}^{H}\sum_{j=1}^{H} g_i g_j E\left\{y_i y_j\right\}. \qquad (5.1.22)$$

Steps involved in deriving the eigenfiltered version of (5.1.20) are identical to the ones discussed for the one given in (5.1.15). Hence, we write the eigenfiltered version of (5.1.20) as

$$\tilde{\sigma}_{\mathrm{a}}^2 = \sum_{k=1}^{L} \lambda_k \left(\sum_{i=1}^{N} q_i \phi_i^{(k)} \sigma_i\right)^2 + \sum_{i=1}^{N} \varepsilon_{ii} q_i^2 \sigma_i^2, \qquad (5.1.23)$$

where λ_k is the kth eigenvalue, L is the number of selected eigenvalues, $\phi_i^{(k)}$ is the ith element of the eigenvector corresponding to the kth eigenvalue, σ_i is the volatility of the ith asset, and ε_{ii} is the error term defined in (5.1.13). We resort to the relative value model theory (Section 3.5) and model the returns of the jth hedging asset as a weighted sum of the investment assets in the portfolio as follows

$$y_j = \sum_{i=1}^{N} \gamma_{j,i} r_i + \xi_j. \tag{5.1.24}$$

In practice, the number of observations is limited to M. Then, we can rewrite (5.1.24) in matrix form as

$$\mathbf{y}_j = \mathbf{R}\boldsymbol{\gamma}_j + \boldsymbol{\xi}_j, \tag{5.1.25}$$

where $\mathbf{y}_j$ is the $M \times 1$ return vector for the jth hedging asset, $\mathbf{R}$ is the $M \times N$ matrix of (investment) asset returns, $\boldsymbol{\xi}_j$ is the $M \times 1$ error vector, and $\boldsymbol{\gamma}_j$ is the $N \times 1$ vector of regression coefficients. By using the eigenanalysis of the empirical correlation matrix of asset returns given in (5.1.9), with $\boldsymbol{\Phi}\boldsymbol{\Phi}^{\mathrm{T}} = \mathbf{I}$ and $\boldsymbol{\Sigma}^{-1}\boldsymbol{\Sigma} = \mathbf{I}$, we rewrite (5.1.25) as

$$\begin{aligned} \mathbf{y}_j &= \mathbf{R}\boldsymbol{\Sigma}^{-1}\boldsymbol{\Phi}\boldsymbol{\Phi}^{\mathrm{T}}\boldsymbol{\Sigma}\boldsymbol{\gamma}_j + \boldsymbol{\xi}_j \\ &= \mathbf{F}\boldsymbol{\beta}_j + \boldsymbol{\xi}_j, \end{aligned} \tag{5.1.26}$$

where $\mathbf{F} \triangleq \mathbf{R}\boldsymbol{\Sigma}^{-1}\boldsymbol{\Phi}$ is the $M \times N$ principal components matrix [28] with its elements F_{nk} being the nth sample value of the kth principal component given as

$$F_k = \sum_{i=1}^{N} \frac{1}{\sigma_i} r_i \phi_i^{(k)}. \tag{5.1.27}$$

We note that the pairwise correlation between two different principal components is zero, and the variance of a particular principal component is equal to its corresponding eigenvalue stated as

$$E\{F_i F_j\} = \begin{cases} \lambda_i & i = j, \\ 0 & i \neq j. \end{cases} \tag{5.1.28}$$

We also note that the principal component given in (5.1.27) is nothing else but the kth eigenportfolio (Section 3.5.4). Now, we can estimate the regression coefficient vector $\boldsymbol{\beta}_j$ given in (5.1.26) by using the least-squares algorithm as follows

$$\hat{\boldsymbol{\beta}}_j = \left(\mathbf{F}^{\mathrm{T}}\mathbf{F}\right)^{-1} \mathbf{F}^{\mathrm{T}}\mathbf{y}_j. \tag{5.1.29}$$

It follows from (5.1.26) that we can regress the return of a hedging asset y_j over L principal components expressed as

$$y_j = \sum_{k=1}^{L} \beta_{j,k} F_k + \zeta_j, \tag{5.1.30}$$

where $\beta_{j,k}$ is the regression coefficient for the jth hedging asset and kth principal component, and ζ_j is the residual term. It follows from (5.1.26) and (5.1.30) that

$$\zeta_j = \sum_{k=L+1}^{N} \beta_{j,k} F_k + \xi_j. \tag{5.1.31}$$

By assuming that the residual term, ζ_j, is orthogonal to all principal components and substituting (5.1.30) in (5.1.22), we write the eigenfiltered version of (5.1.22) as

$$\tilde{\sigma}_{\mathrm{h}}^2 = \sum_{i=1}^{H} \sum_{j=1}^{H} g_i g_j \left(\sum_{k=1}^{L} \sum_{l=1}^{L} \beta_{i,k} \beta_{j,l} E\{F_k F_l\} + E\left\{\zeta_i \zeta_j\right\} \right). \tag{5.1.32}$$

Then, it follows from (5.1.28) that

$$\tilde{\sigma}_{\mathrm{h}}^2 = \sum_{i=1}^{H} \sum_{j=1}^{H} g_i g_j \left(\sum_{k=1}^{L} \lambda_k \beta_{i,k} \beta_{j,k} + E\left\{\zeta_i \zeta_j\right\} \right). \tag{5.1.33}$$

Assuming the cross-correlation between residual terms is zero, $E\left\{\zeta_i \zeta_j\right\} = 0$ for $i \neq j$, we rewrite (5.1.33) as

$$\tilde{\sigma}_{\mathrm{h}}^2 = \sum_{k=1}^{L} \lambda_k \left(\sum_{j=1}^{H} g_j \beta_{j,k} \right)^2 + \sum_{i=1}^{H} g_i^2 \nu_i^2, \tag{5.1.34}$$

where $\nu_i^2 = \mathrm{var}(\zeta_i)$. In a similar fashion, assuming that there is no correlation between asset returns and residual terms, $E\left\{r_i \zeta_j\right\} = 0$, we can express the eigenfiltered version of the cross-correlation between the ith (investment) asset and the jth hedging asset of (5.1.21) using (5.1.30) as follows

$$E\left\{r_i y_j\right\} = \sum_{k=1}^{L} \beta_{j,k} E\{r_i F_k\}. \tag{5.1.35}$$

Substituting (5.1.27) in (5.1.35) yields

$$E\{r_i y_j\} = \sum_{k=1}^{L}\sum_{l=1}^{N} \beta_{j,k}\frac{\phi_l^{(k)}}{\sigma_l}E\{r_i r_l\}. \tag{5.1.36}$$

By substituting (5.1.14) in (5.1.36), due to the orthogonality of the eigenvectors, (5.1.36) becomes

$$E\{r_i y_j\} = \sum_{k=1}^{L} \lambda_k \phi_i^{(k)} \beta_{j,k}\sigma_i. \tag{5.1.37}$$

Hence, the eigenfiltered version of (5.1.21) is obtained as

$$\tilde{\sigma}_{\mathrm{a}}\tilde{\sigma}_{\mathrm{h}} = \sum_{k=1}^{L}\lambda_k \sum_{i=1}^{N}\sum_{j=1}^{H} q_i g_j \phi_i^{(k)}\beta_{j,k}\sigma_i. \tag{5.1.38}$$

Finally, the substitution of (5.1.23), (5.1.34), and (5.1.38) in (5.1.19), and rearranging components leads us to the risk expression, for a hedged portfolio with return given in (5.1.18) by using filtered version of the empirical correlation matrix, as follows

$$\tilde{\sigma}_{\mathrm{p}} = \left[\sum_{k=1}^{L}\lambda_k\left(\sum_{i=1}^{N} q_i\phi_i^{(k)}\sigma_i + \sum_{j=1}^{H} g_j\beta_{j,k}\right)^2 + \sum_{i=1}^{N}\varepsilon_{ii}q_i^2\sigma_i^2 + \sum_{j=1}^{H}\nu_j^2 g_j^2\right]^{1/2}, \tag{5.1.39}$$

where λ_k is the kth eigenvalue, $\phi_i^{(k)}$ is the ith element of the kth eigenvector, σ_i is the volatility of the asset, $\beta_{j,k}$ is the hedging factor for the jth hedging asset and kth principal component, and ν_j^2 is the variance of the idiosyncratic component of the returns for the jth hedging asset regressed on L principal components defined in (5.1.31). See file `risk_hedged.m` for the MATLAB code for the risk calculator given in (5.1.39).

5.2 RISK ESTIMATION FOR TRADING IN MULTIPLE FREQUENCIES

In this section, we continue the discussion on *trading in multiple frequencies* (Section 4.8) by estimating the risk in such a scenario. Here, we highlight the Epps effect [59] stating that cross-correlation of asset returns is reduced as the sampling (trading) frequency increases (Section 6.3). Therefore, pairwise correlations between the assets significantly vary at different sampling frequencies. We display the number of eigenvalues, L, versus

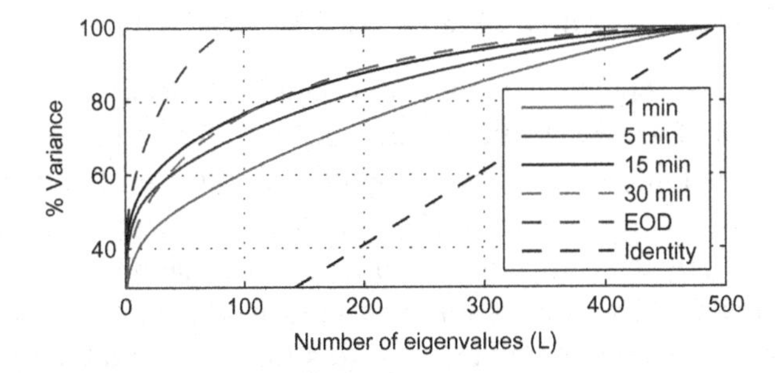

Figure 5.2.3 The number of eigenvalues versus percentage of total variance represented by them for various sampling intervals and the case where $\hat{\mathbf{P}} = \mathbf{I}$, no correlation between assets.

the percentage of the total variance they represent for a typical empirical correlation matrix of asset returns in Figure 5.2.3 with different (trading) sampling frequencies. We see from the figure that eigenspectrum of the correlation matrix becomes more spread as the sampling interval decreases. Thus, more eigenvalues are required to represent a certain percentage of the total variance. EOD (end of day) in the figure stands for price data sampled at the market closing of each day.

In order to accommodate the concept of trading in multiple frequencies, we need to properly modify the risk definition of (3.3.7) to develop a framework. We know from (4.8.1) that

$$\sigma_1 = \sqrt{k_1/k_2}\sigma_2 = m\sigma_2, \tag{5.2.1}$$

where σ_1 and σ_2 are the volatilities estimated at the sampling intervals $T_1 = k_1 T_s$ and $T_2 = k_2 T_s$, respectively, and T_s is the base sampling interval. It is evident from (5.2.1) that we can measure portfolio risk at a certain time interval, and manage risk of assets via trading the individual assets at different time intervals, by expanding the original risk formula of (3.3.7) as follows

$$\sigma_\mathrm{p} = \left(\mathbf{q}^\mathrm{T}\boldsymbol{\Sigma}^\mathrm{T}\mathbf{M}^\mathrm{T}\mathbf{P}\mathbf{M}\boldsymbol{\Sigma}\mathbf{q}\right)^{1/2}, \tag{5.2.2}$$

where $\mathbf{M} = \mathrm{diag}(m_1, m_2, \ldots, m_N)$ and m_i is the scaling factor of (5.2.1) provided that $\boldsymbol{\Sigma}$ and $\mathbf{P}$ matrices are estimated at $T_2 = k_2 T_s$ time intervals. We show the performance improvement achieved by using (5.2.2) instead of (3.3.7) in Section 5.4.4, after discussing the risk management methods (Section 5.4).

5.3 FAST EIGENFILTERING FOR RISK ESTIMATION

Implementation of eigendecomposition is computationally costly. In backtesting (Section 4.4), we iterate through many different scenarios. Hence, this implementation cost can have a significant effect on the overall computational time during backtesting. In this section, we summarize a method introduced in [30] that utilizes a Toeplitz approximation to the empirical correlation matrix. The motivation here is to incorporate the closed-form expressions of eigenvalues and eigenvectors for AR(1) model in the eigenfiltering of correlation matrix to improve the implementation speed. The impact of the approximation error to the risk estimation is shown to be negligible [32]. We start with revisiting the AR(1) signal model. Then we provide our motivation and continue with the definition of two different Toeplitz approximations to the empirical correlation matrix. We finish the section with a discussion on using discrete cosine transform as a replacement to eigenanalysis in filtering.

5.3.1 AR(1) Signal Model

Random processes and information sources of interest in various applications are mathematically described by a variety of popular signal models including auto-regressive (AR), moving average (MA), and auto-regressive moving average (ARMA) types. AR signal model, also called all-pole model, has been successfully used in speech processing for decades [60]. Its first order AR(1) model is a first approximation to many natural signals like images, and has been successfully utilized in many applications. AR(1) signal is generated through the first-order regression formula written as [27, 61]

$$x(n) = \rho x(n-1) + \xi(n), \tag{5.3.1}$$

where $\xi(n)$ is a zero-mean white noise sequence described as

$$\begin{aligned} E\{\xi(n)\} &= 0, \\ E\{\xi(n)\xi(n+k)\} &= \sigma_\xi^2 \delta_k, \end{aligned} \tag{5.3.2}$$

and δ_k is the Kronecker delta function. First-order correlation coefficient, ρ, is real in the range of $-1 < \rho < 1$. In general, mean value of $x(n)$ is assumed to be zero. However, the AR(1) process in certain scenarios is defined as

$$x(n) = \alpha + \rho x(n-1) + \xi(n).$$

For this case, the mean is

$$\mu_x = \frac{\alpha}{1-\rho}, \tag{5.3.3}$$

and the variance of $x(n)$ is given as follows

$$\sigma_x^2 = \frac{1}{(1-\rho^2)}\sigma_\xi^2. \tag{5.3.4}$$

Auto-correlation sequence of $x(n)$ is expressed as

$$R_{xx}(k) = E\{x(n)x(n+k)\} = \sigma_x^2 \rho^{|k|}; \quad k = 0, \pm 1, \pm 2, \ldots. \tag{5.3.5}$$

The resulting auto-correlation matrix of size $N \times N$ for an AR(1) process is of Toeplitz form and shown as

$$\mathbf{R}_x = \sigma_x^2 \begin{bmatrix} 1 & \rho & \rho^2 & \cdots & \rho^{N-1} \\ \rho & 1 & \rho & \cdots & \rho^{N-2} \\ \rho^2 & \rho & 1 & \cdots & \rho^{N-3} \\ \vdots & \vdots & \vdots & \ddots & \vdots \\ \rho^{N-1} & \rho^{N-2} & \rho^{N-3} & \cdots & 1 \end{bmatrix}. \tag{5.3.6}$$

5.3.2 Motivation

Here, we consider a portfolio comprised of 30 stocks in the Dow Jones Industrial Average (DJIA) index along with the index ETF DIA that mimics DJIA (total of $N = 31$ assets). Then, we calculate the empirical correlation matrix $\hat{\mathbf{P}}$ defined in (5.1.4) for these assets using the end of day returns (EOD) for 60 business days, $M = 60$. The time span for the measurement is from March 17 to June 10, 2011. We display the elements of $\hat{\mathbf{P}}$ for each row in a descending order in Figure 5.3.4. The kth sequence in the figure represents pairwise correlations between the kth asset and all other assets in the portfolio. We observe from the figure that an AR(1) source is a good candidate to approximate the sequences given in Figure 5.3.4. Next, we discuss a method to approximate to the empirical correlation matrix.

5.3.3 AR(1) Approximation to Empirical Correlation Matrix

The empirical correlation matrix defined in (5.1.4) is symmetric. We approximate it by a Toeplitz matrix as follows

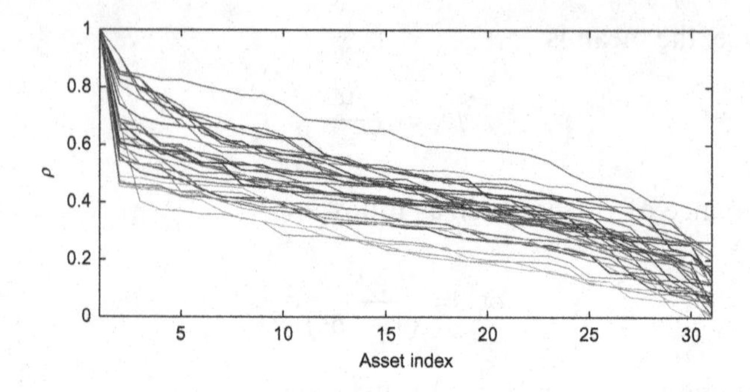

Figure 5.3.4 Rows of $\hat{\mathbf{P}}$ *matrix measured with EOD returns for 30 stocks in index DJIA and index ETF DIA displayed in descending order.*

$$\check{\mathbf{P}} = \begin{bmatrix} 1 & \rho_{\text{opt}} & \cdots & \rho_{\text{opt}}^{N-1} \\ \rho_{\text{opt}} & 1 & \cdots & \rho_{\text{opt}}^{N-2} \\ \vdots & \vdots & \ddots & \vdots \\ \rho_{\text{opt}}^{N-1} & \rho_{\text{opt}}^{N-2} & \cdots & 1 \end{bmatrix}, \tag{5.3.7}$$

where N is the number of assets in the portfolio and ρ_{opt} is the optimal first-order correlation coefficient of AR(1) source found by minimizing the approximation error defined as

$$\varepsilon = \frac{1}{N^2} \sum_{i=1}^{N} \sum_{j=1}^{N} \left(\hat{P}_{ij} - \rho_{\text{opt}}^{|i-j|} \right)^2, \tag{5.3.8}$$

where $\hat{P}_{ij}$ is the element of the empirical correlation matrix located on the ith row and the jth column. Then, we can calculate the resulting eigenvalues and eigenvectors of AR(1) model according to (22) and (25) in [32] as approximations to their measured values, respectively. This approximation speeds up the eigenfiltered estimation of the risk defined in (5.1.15) and (5.1.39). We quantify the approximation error and its impact on the risk estimation of a portfolio in the following example.

Example 5.3. We display the variations of optimum first-order correlation coefficient ρ_{opt} of AR(1) model in time for 30 assets of DJIA and index ETF DIA, $N = 31$, under consideration along with approximation error of (5.3.8) in Figure 5.3.5. We calculate the returns for 24 h intervals with sliding time intervals of 15 min and a measurement window of $M = 60$ business days with a trading day of 6.5 h long. Specifically, we create a

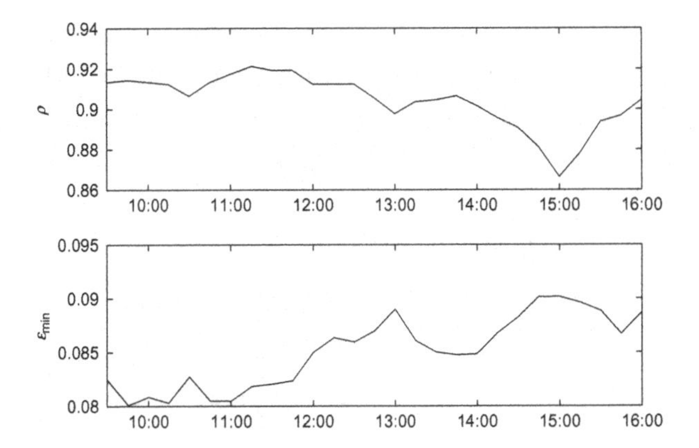

Figure 5.3.5 Variations of optimal first-order correlation coefficient and the resulting error for AR(1) approximation (5.3.8) as a function of time (with 15-min sliding intervals) with $M = 60$ business days for 24-h returns of 31-asset portfolio during trading day 9:30–16:00.

total of 27 return series (one for each 15 min interval of a trading day) of length 60. We calculate each return series by sampling the price series at a specific time on every business day. For example, we calculate the first and last return series by sampling the price at 9:30 and 16:00, respectively, every trading day. Therefore, the last sample in Figure 5.3.5 corresponds to end of day (EOD) return of an asset. Figure 5.3.5 shows highly correlated nature of EOD and 24-h returns.

Furthermore, we can approximate each row of empirical correlation matrix by its own AR(1) signal model, rather than approximating the entire matrix with a single model as before, with the optimum first-order correlation coefficient $\{\rho_{k,\text{opt}}\}$. Hence, we approximate the rows of the empirical correlation matrix as

$$\check{\mathbf{P}} = \begin{bmatrix} 1 & \rho_{1,\text{opt}} & \cdots & \rho_{1,\text{opt}}^{N-1} \\ \rho_{2,\text{opt}} & 1 & \cdots & \rho_{2,\text{opt}}^{N-2} \\ \vdots & \vdots & \ddots & \vdots \\ \rho_{N,\text{opt}}^{N-1} & \rho_{N,\text{opt}}^{N-2} & \cdots & 1 \end{bmatrix}, \tag{5.3.9}$$

where the optimum $\rho_{k,\text{opt}}$ for the kth row of $\check{\mathbf{P}}$ is obtained by minimizing the approximation error as defined

$$\varepsilon_k = \frac{1}{N}\sum_{i=1}^{N}\left(\hat{P}_{ki} - \check{P}_{ki}\right)^2, \tag{5.3.10}$$

and, $\check{P}_{ki}$ is the (k, i)th element of $\check{\mathbf{P}}$ matrix. Then, we approximate each row with an AR(1) source independently. Hence, we rewrite (5.3.9) as

$$\check{\mathbf{P}} = \sum_{k=1}^{N} \mathbf{S}_k \check{\mathbf{P}}_k, \tag{5.3.11}$$

where we define the selection matrix $\mathbf{S}_k$ as

$$\mathbf{S}_k \triangleq \begin{cases} s_{k,k} = 1 & \text{for } k \\ 0 & \text{otherwise} \end{cases}; \quad k = 1, 2, \ldots, N, \tag{5.3.12}$$

and, each $\check{\mathbf{P}}_k$ is a Toeplitz matrix expressed as

$$\check{\mathbf{P}}_k = \begin{bmatrix} 1 & \rho_{k,\text{opt}} & \cdots & \rho_{k,\text{opt}}^{N-1} \\ \rho_{k,\text{opt}} & 1 & \cdots & \rho_{k,\text{opt}}^{N-2} \\ \vdots & \vdots & \ddots & \vdots \\ \rho_{k,\text{opt}}^{N-1} & \rho_{k,\text{opt}}^{N-2} & \cdots & 1 \end{bmatrix}, \tag{5.3.13}$$

for $k = 1, 2, \ldots, N$. We note that it is possible to decompose $\check{\mathbf{P}}_k$ into its eigenvalues and eigenvectors via eigenanalysis as follows

$$\check{\mathbf{P}}_k = \boldsymbol{\phi}_k \boldsymbol{\Lambda}_k \boldsymbol{\phi}_k^{\mathrm{T}}; \quad k = 1, 2, \ldots, N, \tag{5.3.14}$$

where $\boldsymbol{\phi}_k$ is the kth eigenvector. Therefore, we can rewrite the Toeplitz approximation of (5.3.11) as

$$\check{\mathbf{P}} = \sum_{k=1}^{N} \mathbf{S}_k \boldsymbol{\phi}_k \boldsymbol{\Lambda}_k \boldsymbol{\phi}_k^{\mathrm{T}}, \tag{5.3.15}$$

where $\boldsymbol{\phi}_k$ and $\boldsymbol{\Lambda}_k$ are comprised of the kth set of eigenvectors and eigenvalues, respectively, that can be calculated from their closed-form expressions given as (22) and (25) in [32] for the set of AR(1) correlation coefficients $\{\rho_{k,\text{opt}}\}$. We may average the calculated optimum first-order AR(1) coefficients $\{\rho_{k,\text{opt}}\}$ for a simpler implementation as also described in [32].

Example 5.4. For the same basket of Example 5.3, we display the variations of correlation coefficients and resulting approximation errors of this method in Figure 5.3.6. Similar to the previous case, we measured the returns for 24-h intervals with sliding time intervals of 15 min. We note the approximation error of this method is lower than the one for the previous

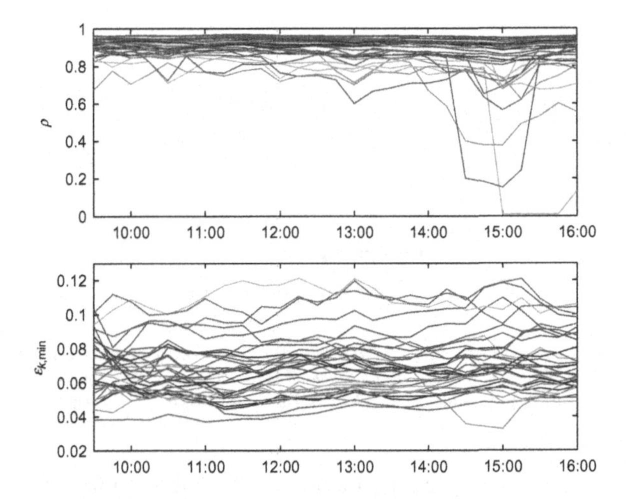

Figure 5.3.6 Variations of optimal correlation coefficients and the resulting errors of AR(1) approximations as a function of time with 15-min sliding intervals for 24-h returns of 31-asset portfolio (30 assets of DJIA index and its index ETF DIA) with $M = 60$ days in the interval 9:30–16:00.

case. The trade-off is the increased computational cost of the multiple Toeplitz approximations.

5.3.4 Portfolio Risk Estimation with Toeplitz Approximation to Empirical Correlation Matrix

In order to test the effect of approximation to the eigenfiltered risk measurement given in (5.1.15), we form a portfolio comprised of 30 stocks in the index DJIA along with its index ETF DIA ($N = 31$). We form the capital allocation vector $\mathbf{q}$ defined in (3.3.5) proportional to the market capitalization of the assets as of June 10, 2011. Elements of $\mathbf{q}$ sum to 1. Similar to the cases in the previous examples in this section, we calculate the empirical correlation matrix $\hat{\mathbf{P}}$ defined in (5.1.4), and its approximations defined in (5.3.7) and (5.3.9) as a function of time with 15-min sliding intervals and a measurement window of $M = 60$ business days for 24-h return intervals. Time span of market data is from March 17 to June 10, 2011. For each case, we keep the largest eigenvalues, $L = 5$ in (5.1.15). For the cases where Toeplitz approximations are used, we used the fast implementations described in [32]. We display the estimation of the eigenfiltered portfolio risk for each case in Figure 5.3.7. The maximum distance in risk estimation from the case where empirical correlation matrix is used is negligible and is equal to +3.67 bps. We observe from Figure 5.3.7

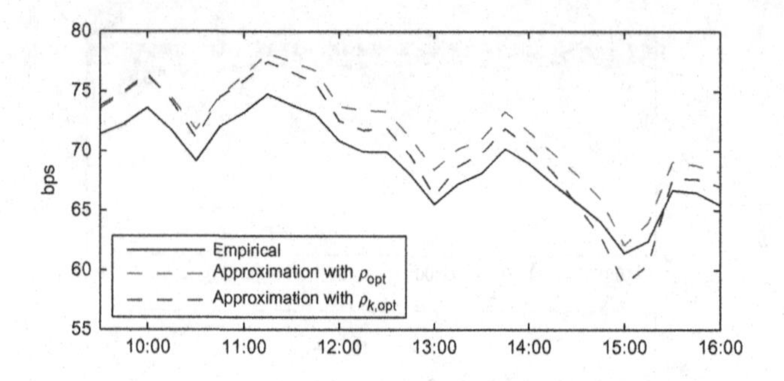

Figure 5.3.7 Variations of portfolio risk calculated from (5.1.15) *with empirical correlation matrix* $\hat{\mathbf{P}}$, *and its Toeplitz approximations* $\check{\mathbf{P}}$ *of* (5.3.7), *and* $\check{\mathbf{P}}$ *of* (5.3.9) *as a function of time with 15-min sliding intervals for 24-h returns of 31-asset portfolio (30 assets of DJIA index and its index ETF DIA) with* $M = 60$ *days in the interval 9:30–16:00.*

that the risk estimations calculated using the Toeplitz approximated, AR(1), empirical correlation matrices have the same proxy to the one where the estimated correlation matrix itself is used.

5.3.5 Noise Filtering with Discrete Cosine Transform

We note that the eigenvectors of the correlation matrix defined in (5.1.10) are nothing else but the basis functions of the *Karhunen-Loève transform* (KLT) [27]. As an approximation to KLT, *discrete cosine transform* (DCT) is preferred due to its ease of implementation in most applications where the first-order correlation coefficient is high [27]. We compare the performances of fixed transform DCT and input signal dependent KLT for empirical correlation matrices of various portfolios in order to justify the use of the former as an efficient replacement to the latter in filtering of the noise in the empirical correlation matrix as given in (5.1.14).

We display the histogram for first-order correlation coefficients of Figure 5.3.6 in Figure 5.3.8. The resulting mean and variance values are 0.8756 and 0.0125, respectively. These results and closeness of KLT and DCT coefficients for the empirical correlation matrix displayed in Figure 5.3.9 suggest the potential use of DCT as a fast KLT approximation in calculating the filtered risk according to (5.1.15) and (5.1.39). The following example quantifies the portfolio risks calculated using eigenfiltering and DCT filtering for noise removal from measured market data, and highlights their comparable performance for the task.

Example 5.5. We employ the identical approach discussed earlier in Section 5.3.4 to test the impact of using DCT as an approximation to KLT in risk estimation. Figure 5.3.10 compares portfolio risks for 24-h returns of

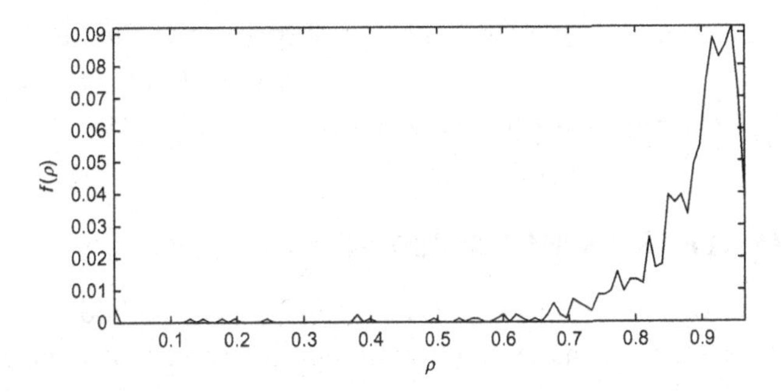

Figure 5.3.8 Histogram of first-order correlation coefficients displayed in Figure 5.3.6.

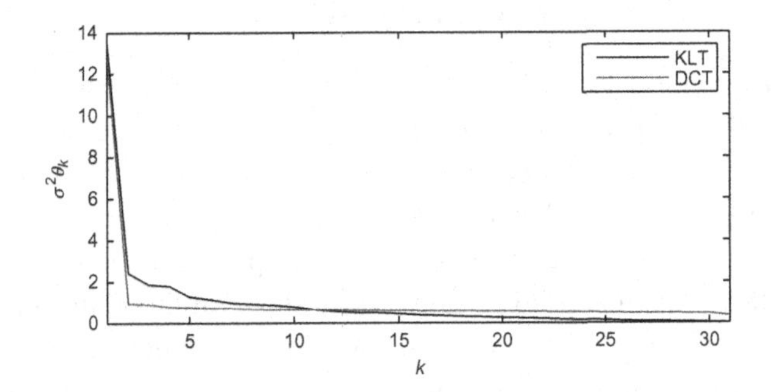

Figure 5.3.9 Coefficient variances of KLT and DCT for $\hat{\mathbf{P}}$ *defined in* (5.1.4).

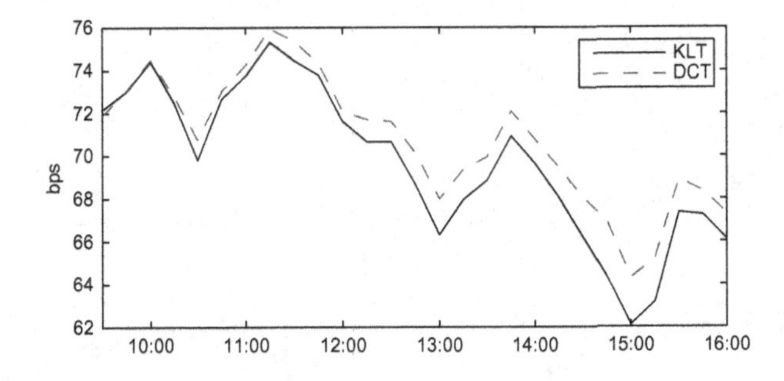

Figure 5.3.10 Portfolio risk calculated from (5.1.15) *using filtered empirical correlation matrix* $\tilde{\mathbf{P}}$ (5.1.14) *as a function of time with 15-min sliding intervals for 24-h return intervals and* $M = 60$ *business days of 31-asset portfolio (30 assets in DJIA index and index ETF DIA) during 9:30–16:00. We performed the noise filtering using KLT basis functions (eigenvectors) and DCT basis functions with* $L = 5$ *and* $L = 10$ *in* (5.1.15), *respectively.*

31-asset portfolio (30 assets in DJIA index and index ETF DIA) calculated from (5.1.15) employing KLT and DCT for noise filtering when five and ten coefficients are kept, $L = 5$ and $L = 10$, as a function of time for 15-min sliding intervals in a given 6.5-h long trading day. Time span is 60 business days from March 17 to June 10, 2011. We observe from the figure that KLT and DCT perform similarly for the noise filtering of empirical correlation matrices for the asset returns experimented.

5.4 PORTFOLIO RISK MANAGEMENT

Once the risk of a portfolio is estimated using the methods presented in this chapter, we can manage it. In some applications, risk management is embedded in the investment strategy itself as in the modern portfolio theory discussed in Section 3.3. However, in trading strategies treated in Chapter 4, risk is not considered. Moreover, in practice, details of the trading strategy might not be necessarily known by the risk manager.

A trivial method for risk management is to manage the portfolio risk by filtering the decisions of the underlying investment strategy based on a predetermined risk limit. We call this method *stay in the ellipsoid* (SIE) since the locus of q_i, $1 \leq i \leq N$ satisfying (3.3.7) for a fixed value of risk σ_p is an ellipsoid centered at the origin [41]. The ellipsoids are nested since as σ_p increases, the ellipsoids become larger. The shape of the ellipsoid is defined by the empirical correlation matrix of asset returns, $\hat{\mathbf{P}}$ defined in (5.1.4). We display the risk ellipsoid for the case of two-asset portfolio in Figure 5.3.11a with $\sigma_p = \sqrt{0.5}$, $\rho_{12} = 0.6$, $\sigma_1 = \sigma_2 = 1$, and $\mathbf{q} \sim \mathcal{N}(\mathbf{0}, \mathbf{I})$. Depending

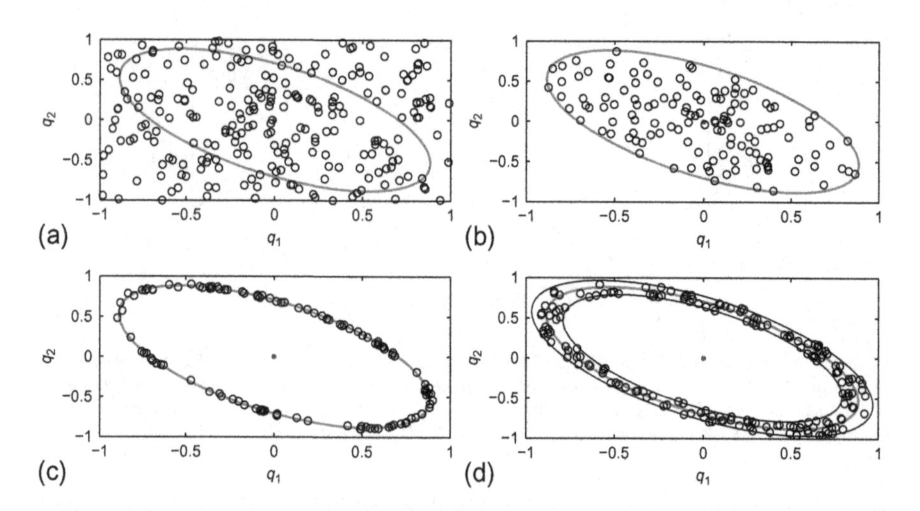

Figure 5.3.11 Possible risk locations of a two-asset portfolio (circles) and the risk ellipsoid for the cases of (a) no risk management, and risk management with (b) stay in the ellipsoid (SIE), (c) stay on the ellipsoid (SOE), and (d) stay around the ellipsoid (SAE) with $\Delta = \sqrt{0.1}$ methods.

on the investment vector $\mathbf{q}$, the risk of a non-managed portfolio, depicted by circles in Figure 5.3.11, may be in, out of, or on the *risk ellipsoid*.

In this section, we present the SIE method along with two of its modified versions that we call *stay on the ellipsoid* (SOE) and *stay around the ellipsoid* (SAE) methods [41]. We note that, in all methods, we assume that once a *signal to enter* a new position is rejected by the risk manager, the underlying strategy does not create another signal until a *signal to exit* is generated.

5.4.1 Stay in the Ellipsoid Method

The goal of stay in the ellipsoid (SIE) method is to keep the portfolio risk anywhere inside the predefined risk ellipsoid by checking the risk of the target portfolio, and rejecting to open any new investment position that violates this requirement. We formulate the SIE risk management method as

$$\mathbf{q}_{t+\Delta t} \leftarrow \begin{cases} \mathbf{q}_{t+\Delta t} & \sigma_{t+\Delta t} < \sigma_{\max}, \\ \mathbf{q}'_{t+\Delta t} & \sigma_{t+\Delta t} \geq \sigma_{\max}, \end{cases} \tag{5.4.1}$$

where Δt is the time interval between the two consecutive portfolio rebalances, $\sigma_{\max}$ is a predetermined maximum allowable risk level, and $\mathbf{q}'_{t+\Delta t}$ is the modified capital investment vector achieved according to the new investment allocation rules

$$\left[q'_{t+\Delta t}\right]_i = \begin{cases} 0 & q_{t,i} = 0 \text{ and } |q_{t+\Delta t,i}| > 0, \\ q_{t+\Delta t,i} & \text{otherwise}, \end{cases} \tag{5.4.2}$$

where $q_{t,i}$ is the investment amount in the ith asset at time t. We note that (5.4.2) employs an *all-or-none* approach in order to expose the portfolio to each asset in approximately equivalent amounts. We observe from (5.4.1) and (5.4.2) that the SIE rejects any new investment position in the target portfolio whenever it generates a target risk higher than the maximum allowable risk. We display the risk locations of possible two-asset portfolios in Figure 5.3.11b along with the limiting risk ellipsoid. Any signal by the underlying trading strategy taking the risk level beyond the target risk outside of the risk ellipsoid is not permitted in the SIE. However, we note that it is still possible for the portfolio to move out of the risk ellipsoid due to the abrupt, unavoidable changes in the returns of the assets that are currently invested. Therefore, other filters like *stop-loss orders* (orders that automatically close a position in an asset if the return on investment is less than a predefined negative value, hence *stopping the loss*) should also be included on the top of the SIE or any other risk management technique employed in practice. See file `sie.m` for the MATLAB code for the implementation of the SIE risk management method.

5.4.2 Stay on the Ellipsoid Method

The *stay on the ellipsoid* (SOE) risk management method aims to keep the portfolio risk not only inside the risk ellipsoid but also as close to it as possible. We observe the difference between the SIE and SOE methods in Figure 5.3.11b and c for the case of a two-asset portfolio. The latter maximizes the utilization of the allowable risk limits. We express SOE as follows

$$\mathbf{q}_{t+\Delta t} \leftarrow \begin{cases} \mathbf{q}_{t+\Delta t} & \sigma_{t+\Delta t} < \sigma_{\mathrm{THR}}, \\ \mathbf{q}'_{t+\Delta t} & \sigma_{t+\Delta t} \geq \sigma_{\mathrm{THR}}, \end{cases} \tag{5.4.3}$$

where σ_{THR} is the desired risk threshold, and $\mathbf{q}'_{t+\Delta t}$ may be obtained by employing various optimization algorithms to minimize the risk distance as expressed

$$\mathbf{q}'_{t+\Delta t} = \underset{\mathbf{q}}{\operatorname{argmin}} |\sigma_{\mathrm{THR}} - \sigma(\mathbf{q})|, \tag{5.4.4}$$

$\sigma(\mathbf{q})$ is the calculated risk for the investment allocation vector $\mathbf{q}$ given that its elements are limited to

$$q_i \in \begin{cases} \{0, q_{t+\Delta t,i}\} & q_{t,i} = 0 \text{ and } |q_{t+\Delta t,i}| > 0, \\ \{q_{t+\Delta t,i}\} & \text{otherwise,} \end{cases} \tag{5.4.5}$$

where the notation $\{\cdot\}$ defines a set of numbers. We note that (5.4.3), (5.4.4), and (5.4.5) suggest to search for a specific combination of signals to open new investment positions targeting the portfolio risk level as close to its limits as possible. The intuition here is to maintain a relatively diverse portfolio while keeping the risk within a desired limit. For the two-asset portfolio case, the solution for the optimization problem is trivial. However, the optimization problem for an N-asset portfolio might become computationally intensive, particularly when N is large.

5.4.3 Stay Around the Ellipsoid Method

The idea behind the *stay around the ellipsoid* (SAE) risk management method is similar to the one for SOE. However, SAE introduces more flexibility by defining a *risk ring* with the help of the two risk ellipsoids located at a fixed distance around the target risk ellipsoid. We observe the difference between SOE and SAE methods in Figure 5.3.11c and d for the two-asset portfolio case. In SAE, it is less likely for a candidate portfolio to be rejected due to the increased flexibility. SAE allows new positions only if the target portfolio risk stays inside the ring. We express SAE as follows

$$\mathbf{q}_{t+\Delta t} \leftarrow \begin{cases} \mathbf{q}_{t+\Delta t} & \sigma_{\min} < \sigma_{t+\Delta t} < \sigma_{\max}, \\ \mathbf{q}'_{t+\Delta t} & \text{otherwise}, \end{cases} \tag{5.4.6}$$

where $\sigma_{\min} = \sigma_{\text{THR}} - \Delta$ and $\sigma_{\max} = \sigma_{\text{THR}} + \Delta$ are the minimum and maximum allowable risk levels defining the risk ring, and Δ is the distance to the target risk. The modified investment vector $\mathbf{q}'_{t+\Delta t}$ in (5.4.6) may be obtained by solving the multi-objective minimization problem defined as [41]

$$\mathbf{q}'_{t+\Delta t} = \underset{\mathbf{q}}{\operatorname{argmin}} \left[f(\mathbf{q}), g(\mathbf{q})\right]^{\mathrm{T}}, \tag{5.4.7}$$

where objective functions f and g are defined as

$$\begin{aligned} f(\mathbf{q}) &= \sigma(\mathbf{q}) - \sigma_{\min}, \\ g(\mathbf{q}) &= \sigma_{\max} - \sigma(\mathbf{q}), \end{aligned} \tag{5.4.8}$$

and $\sigma(\mathbf{q})$ is the calculated risk for investment allocation vector $\mathbf{q}$. Its elements are defined in (5.4.5). The optimization given in (5.4.7) may be performed by creating an aggregate objective function or via various multi-objective optimization algorithms such as successive Pareto optimization [62] or evolutionary algorithms [63].

5.4.4 Performance Comparison of Risk Management Methods

In order to compare the performance of the risk management methods discussed in this section, we perform a backtesting on a portfolio comprised of 100 stocks listed in NASDAQ 100 index as of May 28, 2010. The time span is from April 1, 2010 to May 28, 2010 with the price data sampled at 5-min intervals. We estimate the correlation matrix of asset returns at each sample by using the returns of the past three days, $M = 78 \times 3 = 234$ in (5.1.4). We use the signals from a trading strategy that is known to generate positive P&L in the given time span. At each signal, we invest 4% of the capital in a particular stock, $\delta = 0.04$ in (4.4.5). We consider zero interest rate, $r_{\mathrm{f}} = 0$, and a friction (slippage) of 1.5 bps, $\epsilon = 0.015\%$, for the portfolio equity formula given in (4.4.4).

We display the P&L for the strategy without any risk management in Figure 5.4.12a (solid line). Similarly, we display the P&L curves for the risk managed cases in Figure 5.4.12a for SIE, SOE, and SAE methods. In all methods, risk threshold is set to 3 bps/sample (~25 bps/day). Average

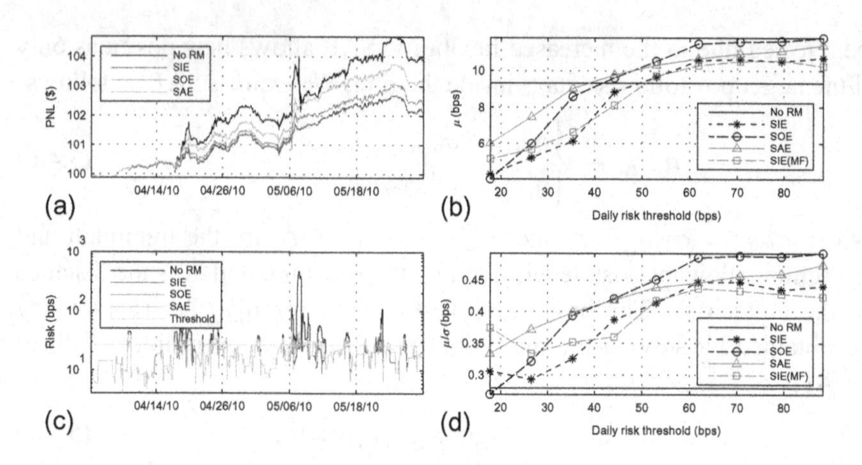

Figure 5.4.12 (a) P&Ls for no risk management case and the scenarios with SIE, SOE, and SAE risk management methods, (b) corresponding estimated daily risk (3.3.7) *values normalized to equity* (4.4.4), σ_P/E, *(c) average daily return versus daily risk threshold for SIE, SOE, SAE, and multiple frequency SIE methods along with the average daily return of no risk management case, and (d) corresponding daily Sharpe ratios.*

daily returns are of 9.4 bps (0.094%), 5.2 bps, 5.9 bps, and 7.4 bps; daily volatilities are of 33.4 bps, 17.8 bps, 18.7 bps, and 20.1 bps; daily Sharpe ratios are of 0.28, 0.29, 0.32, and 0.37; and average numbers of transactions per day are of 25.3, 16.8, 17.8, and 19.7; for no risk management, SIE, SOE, and SAE methods, respectively. We emphasize that the day after the Flash Crash of May 6, 2010 (Section 6.4.4) is interesting since the risk managed strategies avoid the 1.8% draw-down the strategy without any risk management suffered. We display the estimated risk values in Figure 5.4.12b for all the scenarios. We observe from the figures that SAE method outperforms SIE and SOE methods in terms of average return while keeping the volatility at a desired level. All of the methods considered perform well in terms of keeping the portfolio risk bounded with the trade-off of reduced return. However, a less risk-averse investor may easily set the risk threshold to a higher level to increase the level of desired return.

We repeat this experiment by increasing the risk threshold from 2 to 10 bps/sample (from ~17 bps/day to ~88 bps/day). We display the average daily return and daily Sharpe ratio of the strategies for non-managed risk case and for all managed cases in Figure 5.4.12c and d, respectively. We observe from the figures that the SAE and SOE methods yield significantly higher returns with a negligible increase in the volatility than the others for a given risk level.

We also display the P&L performance of the SIE method with the multiple frequency risk estimation formulated in (5.2.1) and (5.2.2) with the sampling rates of $k_1 = 1$, $k_2 = 3$, and $T_s = 5$ min in Figure 5.4.12c and d. with green colored lines. In this scenario, we trade all the assets in the portfolio at the same frequency although the framework introduced in Section 4.8 allows investors to trade different assets at different frequencies. However, we only present the trivial multiple frequency trading results in this section to highlight the flexibility of the framework.

5.5 SUMMARY

Portfolio risk is modeled as a function of pairwise asset return correlations in a portfolio. In other words, portfolio risk is a function of investment allocation vector and empirical correlation matrix of asset returns. One needs to estimate $N(N-1)/2$ pairwise correlations of asset returns to form $N \times N$ empirical correlation matrix $\hat{\mathbf{P}}$ for the portfolio. It is known that $\hat{\mathbf{P}}$ has inherent market noise that may lead to poor estimations of risk in practice. Eigenfiltering is a popular method to filter out this noise. It involves to decompose the matrix $\hat{\mathbf{P}}$ into its eigenvalues and eigenvectors, and sort eigenvalues (and the corresponding eigenvectors) in descending order. Then, reconstruct the matrix by using only the largest $L \ll N$ eigenvalues and corresponding L eigenvectors. The random matrix theory guides us to identify proper value of L to achieve improved performance. It is possible to extend this framework for the case of hedged portfolios where every asset is hedged with a hedging asset by marrying eigenfiltering operator with the relative value theory (Section 3.5). Statistical arbitrage (Section 4.6) is a trading strategy where one needs such an extension. It is also possible to extend the risk estimation for the case of trading in multiple frequencies (Section 4.8) with the assumption that price of each asset in the portfolio is modeled as a geometric Brownian motion process. Since eigenanalysis is costly, one can utilize Toeplitz approximation to empirical correlation matrix of asset returns by using currently available close-form expressions for the eigenanalysis of AR(1) process in order to efficiently estimate portfolio risk. A simple method to manage risk is to reject opening a new position that takes portfolio risk outside of the risk ellipsoid for the allowed risk level. This method is called stay in the ellipsoid (SIE). Its extensions, stay on the ellipsoid (SOE) and stay around the ellipsoid (SAO) better utilize the risk budget and, therefore, yield improved P&L performances over SIE for the underlying trading strategies.

CHAPTER 6

Order Execution and Limit Order Book

So far, we have introduced the financial products (Chapter 2), their price formation and market value (Chapter 3), building trading strategies (Chapter 4), and developing a risk management strategy (Chapter 5). In this chapter, we discuss the execution of the orders to buy and sell assets. In backtesting, we assumed that we can buy and sell any asset for any number of units (even fractional) at any price. In this section, we revisit the problem of market impact and implementation shortfall, the effect of our orders on the market price of an asset. We discuss commonly used execution strategies that naturally lead us to the very fundamental concept of *limit order book* (LOB) in the market microstructure. We explain *limit order* and *market order* types, how they interact with and evolve the LOB of an asset. Then, we overview research on *LOB models*. We get knowledgeable with the inner workings of the market microstructure, then, we discuss the reasons behind the Epps effect, diminishing cross-correlations between asset returns when the sampling (trading) frequency gets higher and higher. Finally, we discuss the category of *high frequency trading* (HFT). We review well known HFT strategies, and discuss the rich literature on covariance estimation with the high frequency market data. We finish the chapter by summarizing the low-latency (ultra high frequency) trading and highlighting the impact of HFT on the financial markets.

6.1 MARKET IMPACT AND ALGORITHMIC TRADING

One of the challenging problems in finance is to execute an order without impacting the market price, hence, not lowering the profit. With small orders, the size of the order is small compared to the overall trading volume

A Primer for Financial Engineering. http://dx.doi.org/10.1016/B978-0-12-801561-2.00006-X

of the asset, impact may be negligible. However, if we place a very large order, market participants can easily detect the high volume we intend to buy or sell, and quickly adjust their offers upward or bids downward as naturally expected, respectively [64–67]. Even if the order size is small, depending on the state of the limit order book, we may end up buying at a very high price or selling at a very low price (Section 6.2.1). In an effort to minimize these order execution risks, players in the market usually divide large orders into smaller pieces and place sub-orders in a systematic way over a period of time. This practice is often referred to as *algorithmic trading*. However, the word "trading" may be confusing since what we do is *of executing* an order according to a decision making *strategy*. The major decisions like what asset, when, how many shares, and what price are made by the *trading strategy* employed. Trading is a more involved process including the mechanism that leads us to decide on buying or selling an asset (we discuss common trading strategies in Chapter 4). Therefore, we prefer to use the terms *algorithmic execution* and *execution strategy* for order execution.

In general, there are three types of execution strategies. Namely, they are the *macro*, *micro*, and *routing*. A macro strategy decides on how many sub-orders the main order is divided into and when those sub-orders are placed. On the other hand, a micro strategy can further divide the sub-orders into multiple micro-orders. It also decides on whether to place a limit order or a market order (Section 6.2.1), and knows what to do when an order is partially filled or when the market moves away from the price at the time sub-order is first placed. A routing strategy decides on which ECN to place the order and whether an order should be re-routed to a different exchange should the price or liquidity for the asset get disadvantageous in the ECN the order was originally placed in. Moreover, a strategy may be *static* or *dynamic*. In a static strategy, we make a decision at the beginning and we follow the strategy accordingly up until the entire order is executed. On the other hand, in a dynamic strategy, we make the decisions on-the-fly based on market conditions and/or the history of the execution up until that point in time.

In this section, we focus on static and macro strategies. We loosely use the terms "placing an order" and "execution" although our macro trader does not necessarily place orders itself through a broker-dealer or an ECN. As discussed above, it is common to have additional layers until execution is completed such as a micro trader and a routing agent in between. However, from macro trader's perspective, the order is placed and executed whenever sent out and execution confirmation with relevant information received.

In the literature, execution is often discussed in the context of getting out of a long position, *liquidating* an asset we own. Although theory is applicable to both getting into and out of a long or short position, we follow the literature in the rest of the section, and assume our goal is to liquidate certain units of an asset that we own. The unit could be shares of stock, contracts of futures, a currency, and others. We start at $t = 0$ and we want to completely liquidate assets at $t = T$ where t is the time variable and T is a predetermined time interval. We sample the time such that there are $N > 0$ sampling intervals in T of length $\tau = T/N$ and $t_k = k\tau$. Let us define the number of units of an asset we own at discrete-time k (continuous time t_k) $x(k)$, also called *execution trajectory*, as

$$x(k) = x(0) - \sum_{i=1}^{k} \Delta x(i) = \sum_{i=k+1}^{N} \Delta x(i), \quad k = 0, 1, ..., N, \tag{6.1.1}$$

where $x(0)$ is what we initially own, and $\Delta x(i) = x(i-1) - x(i)$ is the number of units we liquidate between discrete-times $i - 1$ and i. We expect the price of the asset change as we keep liquidating due to the activity in the market. Initial value of our investment is $s(0)x(0)$ where $s(0)$ is the price of the asset at $t = 0$. Our goal in a macro strategy is to minimize the difference in initial value and value generated through liquidating the asset caused by *total cost of trading* [67] or *implementation shortfall* [68] (or simply *shortfall*) defined as

$$\chi = s(0)x(0) - \sum_{i=1}^{N} s_{\mathrm{E}}(i)\Delta x(i), \tag{6.1.2}$$

where $s_{\mathrm{E}}(i)$ is the execution price for the number of units liquidated $\Delta x(i)$ during the ith trading interval τ_i. Similarly, average execution price is calculated as

$$\bar{s}_{\mathrm{E}} = \frac{1}{x(0)} \sum_{i=1}^{N} s_{\mathrm{E}}(i)\Delta x(i). \tag{6.1.3}$$

Equivalently, our goal is to minimize $s(0) - \bar{s}_{\mathrm{E}}$, namely, we want to liquidate the asset at a price that is close to the initial price. In this setup, the only parameters we can control are the values of $\Delta x(i)$ $\forall i$, T, and N. We highlight that there are two interrelated risks involved as follows.

1. The exogenous activity in the market can move the price far away from $s(0)$ if we wait too long to execute orders. In order to avoid it, we could set $N = 1$ and place an order for the total number of units at once to avoid this risk. However, we may impact the market price if the relative significance of the order volume is high.
2. The market impact moves the price away from $s(0)$ if we place the entire order in one big chunk. In order to avoid it, we could set N (and potentially T) very large such that our movements in the market are negligible in comparison to the overall trading volume. Hence, no market impact is created due to execution of small size orders. However, if we wait too long to complete the execution of the entire order, the price can move far away from $s(0)$.

Therefore, we need a strategy to execute the order such that shortfall is small. It is worthy to note that although it is not our intention when designing a macro strategy, the shortfall (6.1.2) can be negative. That means we liquidate for a value that is larger than the original value. However, this would only prove that we were lucky for that particular execution due to the favorable market conditions for the asset during the execution process.

Example 6.1. Suppose we have 900 shares of a stock and we liquidate it with three sub-orders. The initial price of the stock is \$40 and we execute 200 shares at \$39.94, 400 shares \$39.72, and 300 shares at \$39.80, respectively. Therefore, shortfall in this scenario is calculated using (6.1.2) as

$$\begin{aligned}\chi &= \$40 \cdot 900 - (\$39.94 \cdot 200 + \$39.72 \cdot 400 + \$39.80 \cdot 300)\\ &= \$184,\end{aligned}$$

that is 0.51% loss on the initial value. The average execution price defined in (6.1.3) is calculated as

$$\begin{aligned}\bar{s}_{\mathrm{E}} &= \frac{1}{900}(\$39.94 \cdot 200 + \$39.72 \cdot 400 + \$39.80 \cdot 300)\\ &= \$39.795,\end{aligned}$$

that is similarly 0.51% lower than the initial asset price. Decision on when and for how much to liquidate can have a significant effect on the profit and loss. See file `shortfall.m` for the MATLAB code for an example of this important concept in trading.

6.1.1 Time-Weighted Average Price (TWAP)

The simplest macro strategy is to divide the order into N sub-orders of the same size, $\Delta x(i) = x(0)/N$, and place them at a constant rate. In other words, the number of units liquidated on the interval τ_i is equal to

$$\Delta x(i) = \frac{1}{N}x(0) = \frac{t_i - t_{i-1}}{T} = \frac{\tau_i}{T}x(0), \tag{6.1.4}$$

where T is the total time. Corresponding execution price in the interval τ_i is equal to

$$s_{\mathrm{E}}(i) = s(i), \tag{6.1.5}$$

where $s(i)$ is the average price of the asset between t_{i-1} and t_i. It follows from substitution of (6.1.4) and (6.1.5) into (6.1.3) that the average execution price $\bar{s}_{\mathrm{E}}$ is actually the *time-weighted average price* (TWAP) of an asset defined as

$$s_{\mathrm{TWAP}} = \frac{1}{T}\sum_{i=1}^{N} s(i)\,(t_i - t_{i-1})\,. \tag{6.1.6}$$

Therefore, this macro strategy is called *TWAP*.

Example 6.2. Suppose we observe the price of a stock over the course of an hour ($T = 60\,min$) for every 15-min interval ($\tau = 15\,min$ and $N = 4$) as \$7.12, \$7.86, \$7.45, and \$8.12, respectively. We calculate the TWAP from (6.1.6) as

$$\begin{aligned} s_{\mathrm{TWAP}} &= \frac{1}{60}(\$7.12 \cdot 15 + \$7.86 \cdot 15 + \$7.45 \cdot 15 + \$8.12 \cdot 15) \\ &= \$7.6375. \end{aligned}$$

If we were to liquidate this stock with a TWAP strategy, the average execution price (6.1.3) would be \$7.6375. See file `twap.m` for the MATLAB code for this example.

6.1.2 Volume-Weighted Average Price (VWAP)

The price process of an asset does not necessarily reflect its overall value. A price sample accompanied by a large volume is more informative than the one accompanied by a smaller trading volume. Moreover, it is known that on a usual day, trading volume curve is U-shaped. The volume is the largest close to market open and market close times, and lowest in the middle of the

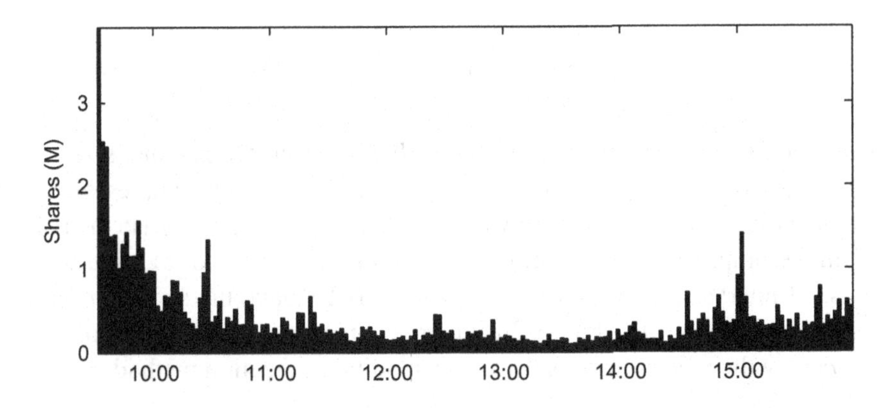

Figure 6.1.1 Trading volume variations of Apple Inc. stock (AAPL) on September 4, 2014.

day. In Figure 6.1.1, we display the intraday trading volume for Apple Inc. (AAPL) stock on September 4, 2014.

A TWAP macro strategy ignores the volume. Thus, it does not necessarily deliver liquidation at a good value (see Example 6.3 for a clarification on "good value"). An alternative strategy is to divide the order into N sub-orders with uneven sizes. We calculate the size of each sub-order according to the ratio of the (expected) trading volume in the time interval τ to the total volume traded in the interval T given as

$$\Delta x(i) = \frac{v(i) - v(i-1)}{V} x(0), \tag{6.1.7}$$

where $v(i-1)$ and $v(i)$ are the cumulative trading volumes at times t_{i-1} and t_i, respectively. And, V is the total trading volume in the interval T defined as

$$V = v(N) - v(0) = \sum_{i=1}^{N} [v(i) - v(i-1)]. \tag{6.1.8}$$

We note that cumulative volume is a monotonically increasing (non-decreasing) process where $v(i) \geq v(i-1)$. Similar to the TWAP, corresponding execution price results in

$$s_{\mathrm{E}}(i) = s(i), \tag{6.1.9}$$

where $s(i)$ is the average price between t_{i-1} and t_i. It follows from substitution of (6.1.7) and (6.1.9) into (6.1.3) that the average execution price $\bar{s}_{\mathrm{E}}$ is actually the *volume-weighted average price* (VWAP) of an asset calculated as

$$s_{\mathrm{VWAP}} = \frac{1}{V}\sum_{i=1}^{N} s(i)\left[v(i) - v(i-1)\right]. \tag{6.1.10}$$

Therefore, this macro strategy is called *VWAP*. We emphasize that sizes of sub-orders with VWAP can vary wildly depending on the samples of the trading volume process. We know the fact that large orders increase the risk of market impact. A modification to the strategy may be not to trade with constant intervals, $\tau_i \neq \tau_j$ for $i \neq j$, where trades have the same constant order size. In other words, we execute at a constant rate on a time domain *warped* by the volume process in such a modified execution method.

Moreover, VWAP requires a good estimation of the volume. We do not know the volume of an interval at its beginning. Assuming that the trading volume over the course of a day will be similar to the previous days, an average volume for the interval across the last L days can be measured. Alternatively, if trader has a good model for the volume process, a prediction of the volume for the upcoming interval may be utilized. We note that the latter is a dynamic macro strategy.

Example 6.3. Suppose we observe the cumulative trading volume in time, $v(n)$, for the stock prices given in Example 6.2 as 0.8K shares at the beginning of the hour and 1K, 1.2K, 2.4K, and 2.5K shares at the end of each 15-min interval. We calculate the VWAP from (6.1.10) as

$$\begin{aligned} s_{\mathrm{VWAP}} &= \frac{1}{1.7}(\$7.12 \cdot 0.2 + \$7.86 \cdot 0.2 + \$7.45 \cdot 1.2 + \$8.12 \cdot 0.1) \\ &= \$7.4988. \end{aligned}$$

If we were to liquidate this stock with a VWAP strategy, the average execution price (6.1.3) would be \$7.4988. It is closer to \$7.45, the price executed with the highest volume. See file `vwap.m` for the MATLAB code for this example.

6.1.3 Optimal Order Execution

So far we revisited the concept of market impact and discussed a couple of intuitive order execution methods to avoid it. The motivation in TWAP and VWAP strategies is to reduce the market-impact by dividing order in sub-orders and execute at constant intervals of time and volume, respectively. A more systematic approach is to use a price model that incorporates the market impact. Then, we can estimate the model parameters and find an optimized strategy that minimizes the expected shortfall (6.1.2) of liquidation for a given risk. This practice is called the *optimal order execution*. In this section, we revisit the celebrated paper on the topic by Almgren and Chriss [67] and summarize the fundamentals of the concept.

6.1.3.1 Market Impact and Implementation Shortfall

We start with modeling the market price of an asset. We modify the log-price model defined in (3.1.24) as [67]

$$s(n) = s(n-1) + \sigma\tau^{1/2}\xi(n) - \tau g\left(\frac{\Delta x(n)}{\tau}\right), \tag{6.1.11}$$

where σ is the volatility of the asset, $\xi(n)$ is a zero-mean unit-variance stationary random process, $g(\cdot)$ is a function describing the *permanent market impact*, $x(n)$ is the number of units of the asset we own at discrete time n, $\Delta x(n) = x(n-1) - x(n)$, and τ is the time we allocate to liquidate $\Delta x(n)$ units. We assume in (6.1.11) that price evolves according to the exogenous factor, volatility (second term), and the endogenous factor, market impact (third term). Similarly, the execution price is modeled as [67]

$$s_{\mathrm{E}}(n) = s(n-1) - h\left(\frac{\Delta x(n)}{\tau}\right), \tag{6.1.12}$$

where $h(\cdot)$ is a function representing the temporary market-impact. We note that effect of $h(\cdot)$ is only visible in the interval τ. It only effects the execution price $s_{\mathrm{E}}(n)$, and not the next market price $s(n)$. From (3.2.13), (6.1.11), and (6.1.12), we can see that shortfall given in (6.1.2) is equal to [67]

$$\chi = \sum_{i=1}^{N}\left[-\sigma\tau^{1/2}\xi(i) + \tau g\left(\frac{\Delta x(i)}{\tau}\right)\right]x(i) + \sum_{i=1}^{N} h\left(\frac{\Delta x(i)}{\tau}\right)\Delta x(i). \tag{6.1.13}$$

The first term on the right hand side of (6.1.13),

$$\sum_{i=1}^{N}\sigma\tau^{1/2}\xi(i)x(i),$$

is the effect of volatility. The second and third terms,

$$\sum_{i=1}^{N}\tau g\left(\frac{\Delta x(i)}{\tau}\right)x(i),$$

and

$$\sum_{i=1}^{N} h\left(\frac{\Delta x(i)}{\tau}\right)\Delta x(i),$$

are the loss in value due to the permanent and temporary market impacts, respectively. Expected value and variance of shortfall are calculated from (6.1.13) as follows [67]

$$\mu_\chi = \sum_{i=1}^{N} \tau g\left(\frac{\Delta x(i)}{\tau}\right) x(i) + \sum_{i=1}^{N} h\left(\frac{\Delta x(i)}{\tau}\right)\Delta x(i),$$

$$\sigma_\chi^2 = \sigma^2 \sum_{i=1}^{N} \tau x(i)^2. \tag{6.1.14}$$

Variance of the shortfall is a function of the volatility of the asset, and expected value of the shortfall is a function of permanent and temporary impact functions, $g(\cdot)$ and $h(\cdot)$, as expected.

6.1.3.2 Linear Market Impact Models

We now choose a linear permanent impact function as defined [67]

$$g\left(\frac{\Delta x(i)}{\tau}\right) = \gamma \frac{\Delta x(i)}{\tau}, \tag{6.1.15}$$

where γ is a constant. We note that with this model, it is assumed that every unit we sell reduces the price per unit by γ. Similarly, we choose a linear temporary impact function as defined [67]

$$h\left(\frac{\Delta x(i)}{\tau}\right) = \epsilon \operatorname{sgn}(\Delta x(i)) + \frac{\eta}{\tau}\Delta x(i), \tag{6.1.16}$$

where sgn is the sign (signum) function, ϵ is the bid-ask spread, and η is a constant that depends on the transient market microstructure. We note that with this model, we assume that every unit we sell costs us $\epsilon + \eta/\tau$. Substitution of (6.1.15) and (6.1.16) in (6.1.14) yields the expected value of the shortfall as [67]

$$\mu_\chi = \frac{1}{2}\gamma x^2(0) + \epsilon \sum_{i=1}^{N} |\Delta x(i)| + \frac{\tilde{\eta}}{\tau} \sum_{i=1}^{N} \Delta x^2(i), \tag{6.1.17}$$

where

$$\tilde{\eta} = \eta - \frac{1}{2}\gamma\tau. \tag{6.1.18}$$

6.1.3.3 Optimization

Our goal in optimal execution is to minimize the expected value of shortfall for a given risk. The risk is directly related to the variance of the shortfall. Therefore, we are interested in finding the *execution trajectory* $x(n)$ that minimizes μ_χ for a given σ_χ^2. We can write the Lagrangian for this problem as [67]

$$L\left(x(i), \lambda\right) = \mu_\chi + \lambda\sigma_\chi^2, \tag{6.1.19}$$

where $\lambda > 0$ is the Lagrangian multiplier. We note that λ has also a practical meaning. Namely, it is the constant that measures the *risk aversion*. The optimum execution trajectory $x^*(n)$ is found by setting the partial derivatives of (6.1.19) to zero. We note that $x^*(n)$ guarantees minimum expected shortfall for a given risk aversion measure, λ. As λ varies, $x^*(n)$ traces out a curve in $\left(\sigma_\chi^2, \mu_\chi\right)$ space referred to as *efficient frontier of optimal execution strategies* [67].

By substituting μ_χ and σ_χ^2 from (6.1.17) and (6.1.14), respectively, assuming that $x(n)$ is monotonically decreasing, $\Delta x(n)$ does not change sign, and $\lambda \geq 0$, we have $L\left(x(i), \lambda\right)$ defined in (6.1.19) as a convex quadratic function of $x(n)$. Therefore, we can find the global minimum by setting the partial derivative to zero as [67]

$$\frac{\partial L\left(x(i), \lambda\right)}{\partial x(n)} = 0,$$

that yields

$$\frac{1}{\tau^2}\left[x\left(j-1\right) - 2x(j) + x(j+1)\right] = \tilde{\kappa}^2 x(j), \tag{6.1.20}$$

where $j = 1, 2, \ldots, N-1$ and

$$\tilde{\kappa}^2 = \frac{\lambda\sigma^2}{\tilde{\eta}}. \tag{6.1.21}$$

General solution to (6.1.20) is a combination of exponentials $e^{\pm\kappa t_j}$ with κ satisfying [67]

$$\frac{2}{\tau^2}\left[\cosh\left(\kappa\tau\right) - 1\right] = \tilde{\kappa}^2, \tag{6.1.22}$$

where $\cosh\left(\cdot\right)$ is the hyperbolic cosine function. We note that from (6.1.18) and (6.1.22) we have, $\tilde{\eta} \to \eta$ and $\tilde{\kappa} \to \kappa$ for $\tau \to 0$. The specific solution to (6.1.20) for $x(N) = 0$ is found as [67]

$$x(j) = x(0)\frac{\sinh\left[\kappa\left(T - t_j\right)\right]}{\sinh\left[\kappa T\right]}, \tag{6.1.23}$$

where $\sinh(\cdot)$ is the hyperbolic sine function. We observe from (6.1.23) that the larger the value of κ the faster we deplete the units. Moreover, the reciprocal of κ, $\theta = 1/\kappa$, may be seen as the "*half-life*" of the execution since it is exactly the amount of time required to liquidate the asset by a factor of e [67].

Example 6.4. Let us assume we own 1 million shares of a stock ($x(0) = 10^6$). We would like to liquidate it over 10 days ($T = 10$ days) by sending an order (to the micro strategy engine which then executes the order over the course of a day in even smaller pieces) every day ($N = 10$). Current stock price is $s(0) = \$60$. Measured annual volatility is 5% ($\sigma = \$0.189/\text{share}/\text{day}^{1/2}$). Bid-ask spread is ¢1 per share ($\epsilon = \$0.01/\text{share}$). Average daily trading volume for the stock is 5 million shares. We use the same rationale given in the example of [67] but with a tighter bid-ask spread of \$0.01. Therefore, we have $\gamma = \$2 \times 10^{-8}/\text{share}^2$ and $\eta = \left(\$2 \times 10^{-7}/\text{share}\right)/(\text{share}/\text{day})$. We choose $\lambda_1 = 10^{-5}/\$$. Hence, from (6.1.21) we calculate the execution rate $\kappa_1 = 1.34/\text{day}$. Similarly, for a more risk-averse trader we have $\lambda_2 = 10^{-6}/\$$, and for the limit we have $\lambda_3 \to 0$ (TWAP) with execution rates calculated as $\kappa_2 = 0.42/\text{day}$ and $\kappa_3 \to 0$, respectively. We calculate the corresponding execution trajectories via (6.1.23) and display their plots in Figure 6.1.2. See file `optimal_execution.m` for the MATLAB code of this example.

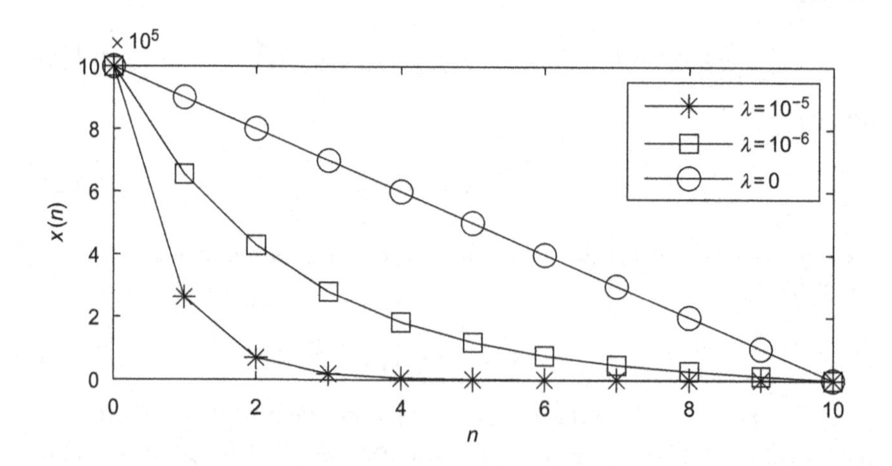

Figure 6.1.2 Optimal execution trajectories defined in (6.1.23) *for different risk aversion measures, λ, for the liquidation of 1 million shares of a stock over 10 days as discussed in Example 6.4.*

6.1.3.4 Further Reading

We refer interested reader to [67] where optimal execution is discussed extensively including the case with non-zero drift, the tradeoff between risk and reward, parameter shifts, liquidating when there are scheduled events and expected news, multiple-security portfolios, and others. Moreover, optimal execution and market-impact are highly active research topics with numerous publications in the literature. For example, see [68–75] and references therein for further details. A paper by Cont [76] and references therein present a discussion on using limit order book models in modeling the market impact.

6.2 LIMIT ORDER BOOK

In the previous section, we identified when and how many units of an asset we want to execute through the macro execution strategy. However, there is much more involved to place and execute an order. In this section, we discuss the *limit order book* (LOB), the structure maintained by ECN in which the *limit orders* are kept and matched against *market orders*. We start with the definition of limit and market orders, and continue with their effect on the evolution of the LOB that facilitates to form market price of an asset. Finally, we discuss analytically tractable models for LOB and their applications. It is desirable to describe the inner workings of LOB through the models where we can analytically express other quantities of interest such as the probability of upward price move based on the current LOB state. We note that understanding LOB helps us not only to build a good execution layer (with increased profit and reduced risk) sitting under the macro execution layer (discussed in the previous section), but also to have better insight to build more complex trading strategies running in much higher frequencies. They are called *high frequency trading* (HFT) strategies as discussed in the next section.

6.2.1 Limit and Market Orders

A *limit order* represents an intention to buy or sell a number assets at the predefined *limit price*. When we place a limit order, we define its size, side, and limit price. For example, a trader can place a limit order to buy (side) 100 shares (size) of the stock AAPL at \$98.75 each (limit price). In this particular case, we say that trader is *bidding to buy* 100 shares of AAPL at \$98.75. Similarly, there are limit orders placed on the other side of the LOB, *asking or offering, to sell* at various quantities and prices. Moreover, there are also many more limit orders in the LOB to sell with varying quantities and prices as well at the same instance. The LOB with its *matching engine* is an *electronic auction system* where its status changes in a microsecond and faster for certain assets. Therefore, at any point in time when the market is open, there may be many traders who are bidding to buy and asking to sell an asset at different prices and various quantities in a typical LOB. The limit order can either be *canceled* or *executed* if it is matched with a market order.

When we place a market order, we define its size and side, and it is executed at the best available price. For example, a trader places a market order to buy (side) 100 shares (size) of the APPL stock. This order is *matched* by the ECN with an existing limit order asking to sell the APPL stock at the lowest price currently available in the market. If the size of that limit order is smaller than the market order being executed, then the remaining shares are matched with the next available limit order that may be asking for either the same or a higher price attached to its quantity. Therefore, a market order to buy may end up buying the asset at a very high price depending on the status of the sell side in the LOB. Therefore, as we discussed in Section 6.1, especially very large orders are executed according to a strategy for a successful *fill*. On the other hand, if the first limit order matched with the market order is larger in size than the size of the market order, they are still matched and market order is successfully filled although the limit order is *partially filled*. A partially filled limit order can still be canceled or matched later with another market order. Similarly, market orders to sell are matched with limit orders bidding to buy the asset.

Assets with large number of limit orders waiting in the LOB to be executed along with large number of trading activity (volume) are called *liquid assets*. Placing limit or market orders are referred to as *adding* or *removing liquidity*, respectively. ECNs, in general, offer a reward or give a penalty for adding and removing liquidity, respectively, to attract liquidity in order to keep the LOB healthy. See Section 4.3 for further details on transaction costs and other specifics of different order types.

Snapshots of an example LOB for a stock are displayed in Figure 6.2.3 before and right after a limit order to buy 100 shares at $98.75 is placed by a trader. Initially, there are five limit orders at four different price levels as shown in Figure 6.2.3a. The two lowest priced limit orders are on the *buy side* while the highest two are on the *sell side*. Therefore, the *bid-ask spread* of the LOB at that instance is $0.01. Then, a limit order to buy 100 shares at $98.75 arrives and placed based on first come first served protocol behind

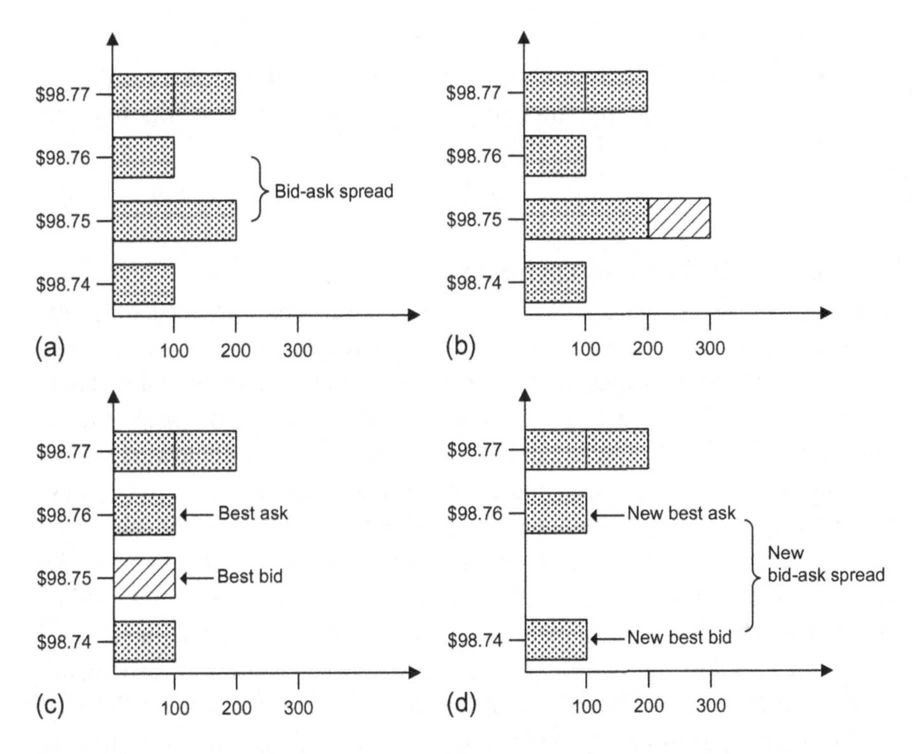

Figure 6.2.3 An example to highlight the evolution of a limit order book in time. (a) Initially, there are five limit orders at four different price levels. The lowest two priced limit orders are on the buy side while the highest two are on the sell side. The bid-ask spread is $0.01. *(b) A limit order to buy 100 shares at $98.75 arrives and placed behind the orders sitting at the same price level. (c) A market order to sell 200 shares at $98.75 arrives and matched with the first arrived limit order sitting at the best bid level. (d) Bidding limit order placed in (b) is canceled. New best bid is $98.74, and the new bid-ask spread is* $0.02.

the orders sitting at the same price level as depicted in Figure 6.2.3b. Later, a market order to sell 200 shares arrives at the LOB and it is matched with the first arrived limit order of 200 shares sitting in the *front* of the *best bid* level. Hence, the available shares at that level of the LOB is depleted by 200 as seen in Figure 6.2.3c. Next, the trader who placed the limit order to buy 100 shares at $98.75 cancels it as displayed in Figure 6.2.3d. Now, the bid-ask spread of the LOB becomes $0.02. Such trading transactions create the supply and demand sides of the market for an asset through its LOB that is *serviced* by *ECN*(*s*). Naturally, the dynamics and stability of the LOB dictate the price volatility (risk) of the asset under consideration where market maker(s), with privileges and responsibilities, play a crucial role. We emphasize the fact that the number of price levels and the number of orders (with quantities attached) in each level (complete book) are much larger in reality than what we have in this simple example (in hundreds for some of liquid stocks). Furthermore, the price level (bin) resolution of LOB and available order types may vary based on asset type and regulatory bodies that set the trading rules, financial and legal frameworks that are implemented and serviced by ECNs.

6.2.2 Levels in the Limit Order Book

LOB has two sides, namely, *bid side* and *ask side*, where limit orders to buy (bidding orders) and limit orders to sell (asking/offering orders) are kept and continuously updated to create and maintain a market for a tradable asset. As displayed in Figure 6.2.3b, a limit order is placed in the bid side of LOB since it is a buy order. The highest bidding (buy side) and the lowest asking (sell side) prices at any moment are called the *best bid* and *best ask prices*, respectively. The price difference between the best ask and bid is referred to as *bid-ask spread*. The larger the bid-ask spread the costlier it is to *cross the spread* to execute a buy or sell order at the best ask price or at the best bid price, respectively. In other words, it is more desirable to *make the spread* when one executes a buy order at the best bid price. There are participants in the market that place large limit orders at both best ask and best bid prices and seek profit from the bid-ask spread. They are called *market makers* as discussed in Section 2.1.

The best bid and best ask prices form the first *level* of bid and ask sides of the LOB, respectively. Similarly, the second (or the *k*th) highest and lowest prices of bidding and asking limit orders form the second (or *k*th) level of the bid and ask sides of the LOB, respectively. Therefore, at any point in time, LOB of an asset in an ECN consists of two sides with multiple levels. In Figure 6.2.3a, we display the first two levels for each side of the LOB. If an asset is traded in multiple ECNs, then, there are multiple LOBs for the same asset at any given time. Naturally, best bid and best ask prices in each LOB may be (and usually are) different. The highest best bid and lowest best ask prices for all the LOBs of an asset maintained by their corresponding ECNs are called as the *national best bid and offer*, and abbreviated as *NBBO*.

The best bid and best ask prices, the first levels for sell and buy sides of LOB, respectively, are referred to as *Level 1* (L1). The first five levels on bid and ask sides of the LOB are referred to as *Level 2* (L2). All the levels on both sides of the LOB are referred to as the *complete book*. For most stocks, the most activity (placement and cancellation of orders) occurs in L2 and the highest trading activity is in L1 as expected [76].

6.2.3 Order Matching and Transactions

Market orders are matched with limit orders by a *matching engine* that is operated by the ECN. Assuming the LOB state is the one displayed in Figure 6.2.3b, a market order to sell 200 shares of APPL stock would be matched with the limit order sitting at $98.75 on the bid side of the LOB. The new state of the LOB after market order executed is displayed in Figure 6.2.3c. When a market order is matched with a limit order, it is a *transaction*, and the resulting *trade price* and *quantity* are recorded. This price is sometimes referred to as the *print* or the *tick*. In general, trader does not specify any specific ECN when placing an order, and market orders are routed by the brokers on behalf of the traders. Brokers are required to route the market order to the ECN with the best bid or ask prices of the NBBO at the moment for its execution.

The limit order placed in the LOB of example, displayed in Figure 6.2.3a as its initial state, transforms it to the state shown in Figure 6.2.3b where it still waits to be filled. This is due to the fact that it is placed *behind* the other limit orders (in our example there is only one) arrived earlier and waiting to be matched at the same level, $98.75, of the LOB. In stock markets, in general, limit orders are matched in the order they are placed in the LOB, first come first served. In some other markets, e.g., EUREX, the orders are matched against all the limit orders waiting at the same level proportional to the size of the orders [77].

A limit order can be canceled. Hence, it can be removed from its corresponding level in the LOB by the trader at anytime. For many liquid stocks in NYSE and NASDAQ, up to 80% (sometimes more) of the limit orders are canceled within the first second they are placed [76] in the LOB. In Figure 6.2.3d, we display the state of the LOB right after the limit order for 100 shares at $98.75 waiting to be matched is canceled. Since there are no bidding limit orders waiting to be filled at $98.75, the new best bid is at $98.74, and the new bid-ask spread is increased to $0.02 from $0.01. A confirmation is sent to the trader by the ECN through the broker after the order is removed from the LOB. However, there might bc a market order ahead in the order flow that is bound to be matched with the limit order, creating a *race condition*. Therefore, canceled but not yet confirmed limit orders can be filled or partially filled. This fact adds extra complexity to the LOB models as well as automated order execution systems.

ECNs stream the order flow and the transactions in real-time to traders for a fee. Users of these real-time data streams are usually the high frequency trading companies. The historical record of the order flow and transactions data are referred to as *quote* and *trade data*, respectively. Practitioners and researchers can subscribe to these data services and download the data for their market analysis. The most widely known and cited quote and trade data store example is the *Trade and Quote Database* (*TAQ*) maintained by New York Stock Exchange (NYSE).

6.2.4 Limit Order Book Models

It is known that the state of the LOB at any point in time conveys information about the short-term price (supply-demand) behavior of the asset [77, 78]. Therefore, it is of great interest in finance to model the LOB. Using the model we can measure statistical indicators and quantities such as the distribution of execution time, distribution of the duration (interval) between price changes, the probability of being able to execute at the best bid or best ask, distribution and auto-correlation of price movements conditional on the state of the LOB, and many others [76]. However, there are additional parameters of the LOB state such as the size at each level and inter-level relationships and dynamics. Moreover, evolution of the LOB is quite complex as order and transaction flows include placement and cancellation of many limit orders in multiple levels with their quantities, and their matching with the incoming market orders. Therefore, building analytically tractable models of LOB is a major challenge in particular for *high frequency trading (HFT)* [76]. In this section, we briefly discuss the multi- and single-level queueing models for LOB. We refer readers of more interest to several publications for further details in the next section.

It follows from our discussion in Section 6.2.3 and corresponding Figure 6.2.3 that a natural choice for an LOB model is a multi-level queue [77, 79–81]. As limit orders arrive, they fill up the queues at the corresponding price levels or bins of the LOB. Queues are depleted either by a matching market order or by cancellations. A Markovian queueing model with the assumption of independent Poisson processes and constant arrival rate, λ, for each level of the LOB is studied in [77]. However, it is observed in practice that arrival rate of orders is not constant, and in contrast, they are clustered in time [80]. It is observed from market data that there is a significant auto-correlation among time durations of order clusters, and positive cross-correlation between the order arrival rates of different order types [76]. LOB models employing Hawkes process [82], self-exciting point-processes with time varying arrival rate $\lambda(t)$, that depend on the history of the order flow for the arrivals of different order types are also proposed in the literature [80]. A simplified queue model for the LOB where only the first level, best bid and best ask prices and the corresponding sizes, is considered in [81]. They discuss and empirically measure *hitting time* that quantifies the time it takes for price at best bid or best ask to change due to a complete depletion. In between different hitting times, they use a Markovian model for the queues with Poisson order arrival to analytically express the distribution of the price changes and inter-change (price) durations as well as to measure the probability of an upward price move.

6.2.5 Further Reading on LOB

The speed and volume of electronic trading is increasing every day in particular with the wide use of high frequency trading (Section 6.4). Studying LOB models and their ties to the price behavior of an asset is the key to having a better understanding of the market microstructure and its dynamics. It is a highly active research area with many applications in finance where the findings are not necessarily disseminated quickly due to the nature of work. We refer interested readers to [83–85] and references therein for further discussions dealing with empirical observations on LOBs, and market microstructure in general. Also, see [72, 86, 87] for applications of LOB models in optimal order execution (Section 6.1.3). Moreover, see references [88, 89] for applications in high frequency trading. Furthermore, an extensive survey is provided on modeling of high frequency financial data in [76].

6.3 EPPS EFFECT

As we discussed in the earlier chapters, a good estimation of correlation among asset returns is crucial for good performance in almost all trading strategies and risk management systems [41]. However, a good correlation estimation and its interpretation, especially in intra-day and high-frequency trading where sampling periods are typically below a minute if not a milli or microsecond, is of a major challenge as also reported in [90–92]. It is well-known in finance that the correlations among financial asset returns decrease as the sampling period of prices decreases. In this section, we revisit this phenomenon called Epps effect [59] and explain its impact in practice.

6.3.1 Trading Frequency (Sampling Period) and Cross-Correlation of Asset Returns

Using (3.2.2) and assuming that the *log-return* expected values of two assets are zero, $\mu_{T_1} = \mu_{T_2} = 0$, we can write their cross-correlation given in (3.2.6) as a function of the sampling period T_s as follows

$$\rho_{12}(T_s) = \frac{E\left\{g_{1,T_s}(n) g_{2,T_s}(n)\right\}}{\sigma_{1,T_s}\sigma_{2,T_s}}, \tag{6.3.1}$$

where $g_{1,T_s}(n)$ and $g_{2,T_s}(n)$ are the log-returns of the first and second assets sampled with T_s, respectively, and the resulting standard deviations, σ_{1,T_s} and σ_{2,T_s}. Similarly, we calculate the cross-correlation of two asset log-returns both sampled with the period $T_s = T_2$ as

$$\rho_{12}(T_2) = \frac{E\left\{g_{1,T_2}(n) g_{2,T_2}(n)\right\}}{\sigma_{1,T_2}\sigma_{2,T_2}}, \tag{6.3.2}$$

where $T_2 = kT_1$. Then, it follows from (3.2.14) and (6.3.2) that

$$\rho_{12}(T_2) = \frac{1}{\sigma_{1,T_2}\sigma_{2,T_2}} E\left\{\sum_{i=0}^{k-1} g_{1,T_1}(kn-i) \sum_{j=0}^{k-1} g_{2,T_1}(kn-j)\right\}. \tag{6.3.3}$$

Let us assume that the cross-correlation of two asset log-returns sampled at two different sampling points in time is zero,

$$E\left\{g_{1,T_s}(n-k) g_{2,T_s}(n-l)\right\} = \rho_{12}(T_s)\,\sigma_{1,T_s}\sigma_{2,T_s}\delta_{k-l}. \tag{6.3.4}$$

From (3.1.26), (3.2.15), (6.3.3), and (6.3.4) we conclude that

$$\begin{aligned}\rho_{12}(T_2) &= \frac{k}{\sigma_{1,T_2}\sigma_{2,T_2}} E\left\{g_{1,T_1}(n) g_{2,T_1}(n)\right\} \\ &= \frac{k}{\sqrt{k}\sigma_{1,T_1}\sqrt{k}\sigma_{2,T_1}} E\left\{g_{1,T_1}(n) g_{2,T_1}(n)\right\} \\ &= \rho_{12}(T_1).\end{aligned} \tag{6.3.5}$$

We show in (6.3.5) that the cross-correlation coefficient, ρ_{12}, between the returns of two assets that follow geometric Brownian motion paths with pure Gaussian increments is independent of the sampling period. Epps [59] was the first to show that the empirical data does not comply with (6.3.5). Moreover, Epps stated that the cross-correlation between two financial asset returns decreases as the sampling period gets smaller,

$$\rho_{12}(T_s) \to 0 \text{ as } T_s \to 0. \tag{6.3.6}$$

6.3.2 Empirical Evidence on Epps Effect

We estimate the cross-correlation coefficient between the log-returns of Apple Inc. stock (AAPL) and PowerShares QQQ Trust ETF (QQQ) using 60 days of historical data between April 1, 2010 and June 30, 2010 as a function of sampling period T_s and display it in Figure 6.3.4a. The sample correlation estimator employed for Figure 6.3.4 is given as

$$\hat{\rho}\left(T_s\right) = \frac{1}{N-1}\sum_{i=0}^{N} \bar{g}_{1,T_s}(n-i)\bar{g}_{2,T_s}(n-i), \tag{6.3.7}$$

where $\bar{g}_{k,T_s}(n)$ is the normalized log-return of the kth asset with zero mean and unit variance as calculated

$$\bar{g}_{k,T_s}(n) = \frac{g_{k,T_s}(n) - \hat{\mu}_k\left(T_s\right)}{\hat{\sigma}_k\left(T_s\right)}, \tag{6.3.8}$$

$\hat{\mu}_k\left(T_s\right)$ and $\hat{\sigma}_k\left(T_s\right)$ are the estimated mean and standard deviation of $g_{k,T_s}(n)$ defined in (3.2.20) and (3.2.19), respectively. Since AAPL is a significant stock of NASDAQ100 index and QQQ mimics the behavior of NASDAQ100 index, we expect to have a significant correlation between the returns of these two relevant assets. However, Figure 6.3.4a suggests a more complicated information. We observe from the figure that the result given in (6.3.5) does not always hold, and the geometric Brownian motion model for the log-price of the assets given in (3.1.25) needs to be improved.

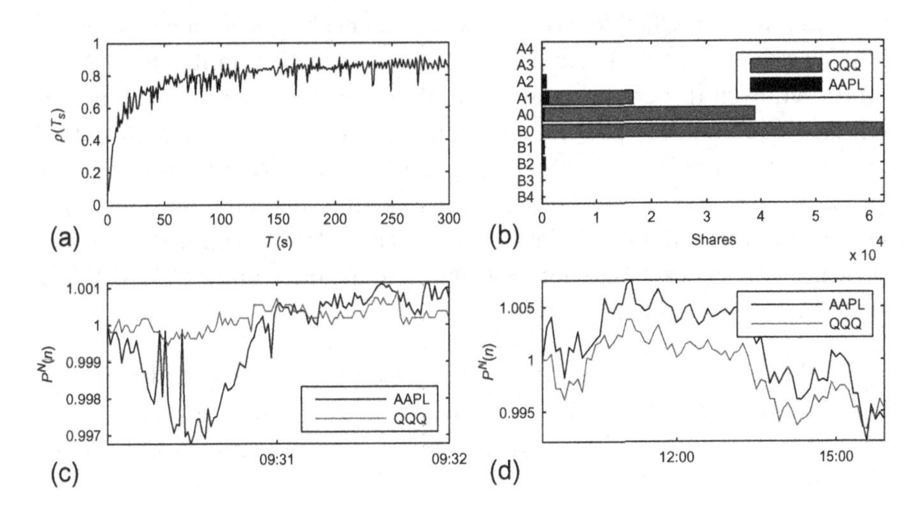

Figure 6.3.4 (a) Cross-correlation between the log-returns of AAPL and QQQ as a function of sampling period T_s, (b) A typical snapshot of the first five levels of the LOBs for AAPL and QQQ, normalized last traded prices of both stocks on March 17, 2011 (c) between 9:30am and 9:32am sampled with $T_s = 1\,s$, (d) between 9:30am and 4:00pm sampled with $T_s = 300\,s$.

This concern has been another active research topic where several authors proposed improved models. See [90, 91] and references therein for details.

One of the culprits for Epps effect is the asynchronous nature of trading. Although prices of the assets within the same industry tend to behave similarly, and they respond to various intra-day economical, social, and political news and data in the same way, they are not traded at the same time points or at the same side of the limit order book (LOB) that is specific for each asset and trading venue. When a trader places a limit order to buy/sell an asset at some specific price and quantity, shares associated with the order are placed in the corresponding price level in the bid/ask side of the LOB. The shares waiting to be bought/sold at the highest/cheapest prices rest in the best bid (B0) and best ask (A0) levels (bins) of the LOB. Traders who place market orders to buy/sell an asset are matched with the shares waiting longest in the ask/bid side starting from the best ask/best bid levels. Moreover, even though two different assets were traded at the same time, their market structures (LOBs), volume, and liquidity, are different and they play a significant role in price formation process. As an example, we display a typical snapshot of the first five levels (on both bid and ask sides) for the LOBs of AAPL and QQQ in Figure 6.3.4b. We observe from the figure that the levels of the LOB for AAPL do not offer too many shares available for selling or buying at that point in time. However, there are many QQQ shares resting at best bid (B0) and best ask (A0) prices at the same time. Therefore, it is less likely for QQQ to have the resting shares depleted on one side and provide a different last-traded price print in short term than it is for AAPL. Moreover, even if there were only two players in the market, who buy and sell the same number of AAPL and QQQ shares, there is no guarantee that they execute the trades synchronously, i.e., first buy AAPL and then sell QQQ or vice-versa. If both players complete the trade in T seconds, then for an observer sampling the last-traded prices with the period $T_s > T$, the asynchronous trades would not be visible, and the price prints of AAPL and QQQ would seem to happen together.

We display the normalized last-traded prices, $p^N(n) = p(n)/p(0)$, of both stocks between 9:30am and 9:32am with $T_S = 1$ s and between 9:30am and 4:00pm with $T_S = 300$ s on March 17, 2011 in Figure 6.3.4c and d, respectively. We observe from these figures that the good proxy between the prices of two stocks that exists at lower sampling rates disappears as the sampling rate increases. This is a very important phenomenon and a serious concern in particular for high frequency trading since the traditional

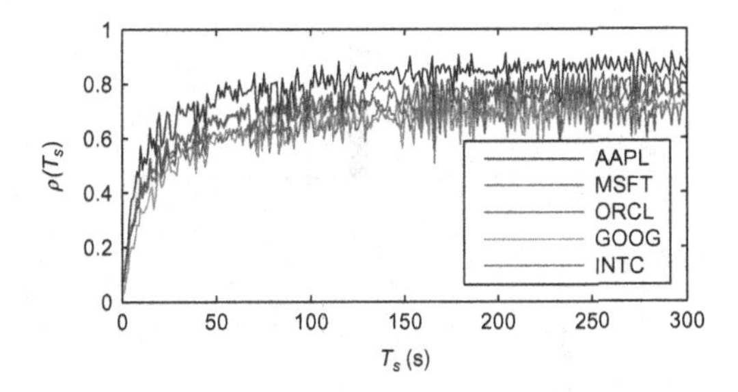

Figure 6.3.5 Cross-correlation between the QQQ (index ETF) and its five largest holdings as a function of sampling period, T_S.

risk management framework does not hold any more to provide practical solutions in order to maintain an investment portfolio rebalanced at short time intervals (high frequencies).

Holdings of QQQ are comprised of NASDAQ 100 technology stocks with their relevant investment factors. Hence, QQQ and those stocks are expected to be correlated. We estimate the correlation coefficients between the log-returns of QQQ (index ETF) and its largest five holdings (AAPL: Apple Inc., MSFT: Microsoft Corp., ORCL: Oracle Corp., GOOG: Google Inc., INTC: Intel Corp.) using 60 days of historical data between April 1, 2010 and June 30, 2010 and display them in Figure 6.3.5 as a function of the sampling period T_S. We observe from these figures that the correlation pattern and Epps effect observed in Figure 6.3.4a. are not only specific to AAPL and QQQ pair.

6.3.3 Trading Frequency and Performance of Sample Estimator

The built-in asynchronicity affects correlation estimation given in (6.3.7) since even the price of one asset of the pair does not change, the product of two returns becomes zero. And, that zero-product term is considered in the averaging operator in the estimator given in (6.3.7). Similarly, in a scenario where the constant prices of both assets traded synchronously are also considered as perfectly uncorrelated pair of returns within this framework. Although these two zero-product cases for correlation calculations are distinct, it is more likely not to have price changes at smaller time intervals. It is reasonable to expect higher correlation if correlation calculation only considers nonzero products with irregular sampling grid where price variations occur for both assets.

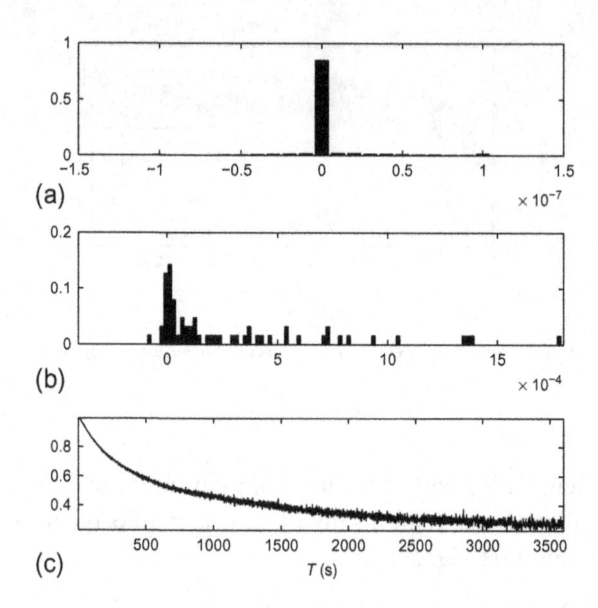

Figure 6.3.6 Histogram of pairwise products for the log-returns of AAPL and QQQ with sampling intervals (a) $T_s = 1\,\text{s}$, and (b) $T_s = 24\,\text{h}$ (EOD). (c) Probabilities of pairwise product terms being negligible, between $-\varepsilon$ and ε for $\varepsilon = 3 \times 10^{-6}$, as a function of sampling interval T_S.

We display in Figure 6.3.6a the histogram of pairwise products for log-returns in the correlation calculation of AAPL and QQQ pair according to (6.3.7) at $T_s = 1$ s sampling period. Similarly, we display in Figure 6.3.6b the histogram for the case of end-of-day (EOD) correlation of the same pair. Figure 6.3.6c depicts probabilities of product terms used in the correlation calculations being negligible, between $-\varepsilon$ and ε for $\varepsilon = 3 \times 10^{-6}$, as a function of sampling period T_S. We note that the probability of having negligible product term in correlation calculation drops when sampling interval increases as highlighted in this figure. This fact has direct impact on the values of pairwise correlations calculated through averaging that also includes negligible ones. Hence, pairwise products for log-returns drop significantly at higher sampling frequencies. We revisit this phenomenon in Section 6.4.2. There is a rich literature on better estimators that leverage high frequency financial data with small values of T_S.

6.4 HIGH FREQUENCY TRADING

High frequency trading (*HFT*) is implemented through sophisticated computer algorithms that constantly seek for opportunities in the market and act on them based on a *trading strategy*. HFT firms heavily invest in technology infrastructure and pay premium fees to have real-time low latency high frequency financial data as well as *direct market access* (DMA). HFT practitioners do not seek large profits per trade. Instead, they go in and out of positions very rapidly (sometimes with holding times that are less than even a microsecond) with very small profits. However, their profits can potentially sum up to very large amounts due to the high number of trades they do over the course of a day. By 2010, more than 60% of all the U.S. equity trading activity was generated by HFT. Though the volume has gotten lower since then,[1] HFT is still a very large contributor of the trading volume around the globe.

6.4.1 HFT Strategies

The literature on HFT strategies is not rich since most of the studies are proprietary and are not made public and high-frequency financial data has not been widely accessible by academics until very recently. We discuss only the well-known strategies here, recognizing the fact that there are many others that are not known by anyone but their inventors and their financial firms.

Market making is one of the most common HFT strategies. Market makers place limit orders (quote) *simultaneously* both on the bid and ask sides of the limit order book (LOB). They seek to make profit from the bid-ask spread. Traditionally, dealers and specialist firms used to be the market makers. However, with the advances in the availability of the electronic markets and venues, there are many market makers with various trading volumes in today's markets. The main difference between a traditional and an HFT market maker is that the latter has no obligation to provide liquidity for the market. Moreover, HFT market makers have access to real-time market data (quotes, trades, all the levels of LOB), use the data to continuously price their bids and asks using mathematical models, and leverage technology to do those across many markets and financial assets within a fraction of a second. Market makers face inventory risk due to the uncertainty of the asset price and asymmetric information risk due to the presence of informed traders in the market. These two risks and development of optimal pricing for market making that minimizes them are studied in the literature [88, 89, 93–97]. The statistical properties of the LOB

[1] See, e.g., Bloomberg Business Week article on June 6, 2013 titled "How the Robots Lost: High-Frequency Trading's Rise and Fall."

such as the arrival rates of buy and sell orders, size distribution of market orders, and the temporary price impact of market orders are reported in [88] where the approach presented in [93] is extended to derive the optimal bid and ask quotes (in the sense that inventory risk is minimized) for the HFT market maker. That framework led to the relevant studies reported in [89, 97]. The study reported in [97] starts from the same model given in [88] and adds inventory limits. Using a new change of variables, they convert the Hamilton-Jacobi-Bellman equations related to the problem into a system of linear ordinary differential equations. Their contribution not only allows simplified computations but also lets to study the asymptotic behavior of the optimal bid and ask quotes. A stochastic framework to model the LOB dynamics is proposed in [77]. Using that model, it is possible to measure several probabilities given the state of the book such as the probabilities of an increase in the mid price, execution of an order at the bid before the ask quote moves, and execution of both a buy and a sell order at the best bid and best ask price, respectively, before the mid price moves. Given the measured probabilities they have, the authors propose an HFT trading strategy that simultaneously places orders on both sides of the LOB betting against mid-price move prior to executions of both orders. If the measured probability is high, the strategy has high confidence that it will make the spread to profit from the difference between the ask and bid prices. Another type of HFT strategy targets to earn the liquidity rebate (Section 4.3) from the ECNs rather than seeking profit from the bid ask spread as market making does.

Some of the HFT strategies are developed around analyzing the real-time quotes data and *detecting movements of institutional players* such as pension funds and mutual funds. Assume a large buy program on a specific stock is detected. Then, the straightforward move would be to buy the same stock early to benefit from potential increase in price due to the large demand expected during the day. Traditionally, if the order is large in size or in dollar value, it is considered as initiated by an institutional player. A discussion on an alternative approach based on regressing quarterly changes in institutional ownership of stocks on cumulative quarterly trades is presented in [98]. Detecting a large order has become a nontrivial problem with a substantial research on optimal order execution methods (Section 6.1.3) and financial firms being specialized in execution of large orders to avoid creating market impact. Recent studies for optimal order execution leverage high frequency data and availability of LOB state through various methods including convex optimization [99], machine learning [100–105], and market-impact models [72, 75].

Another class of algorithms continuously *analyze the news* on a particular company and make trading decisions accordingly. Real-time and historical financial news are available through various electronic feeds and databases. Computers can analyze the news and act much faster than humans. There is evidence on the correlation between prices of financial assets and news as reported in [106]. Numerous methods to analyze news such as textual analysis, mood analysis, fuzzy neural networks obtained from various sources such as Twitter feeds [107], Dow Jones Newswire [108], newspaper headlines [109], and others have been used. A research study measuring the effects of different news sources on the prediction of market volatility and gold prices is reported in [110]. A discussion on the correlation between news and high frequency market reactions is presented in [106].

There are HFT *arbitrage* strategies that look for inefficiencies between the returns of an asset and its derivative (or another highly correlated asset), like a stock and an option on the same stock or a stock and its index ETF, or between the prices of the same asset traded at different venues at the same time, like APPL trading at $100.03 at NYSE and $100.01 at NASDAQ. The former is simply the pairs trading implemented at a higher frequency and the latter is referred to as *exchange arbitrage* [111]. As we discussed in Sections 4.5 and 4.6, pairs trading and statistical arbitrage require good measurement of covariance among assets in a basket. There is a vast literature on how to use high frequency data to estimate volatility and covariance of asset returns. This topic is covered in the next section. Exchange arbitrage and similar low complexity and high speed strategies require almost continuous upgrade in technology since they are only possible through *low-latency trading* (Section 6.4.3).

6.4.2 Covariance Estimation with High Frequency Data

Covariance estimation for asset returns in a basket plays a crucial role in portfolio optimization (Section 3.3.2), factor models (Section 3.5), pairs trading (Section 4.5), statistical arbitrage (Section 4.6), and other applications in finance. The speed of electronic trading is increasing along with more availability of the high frequency market data. However, as we discuss in Sections 6.3.2 and 6.3.3, sampling asset returns faster does not offer valuable covariance information due to the Epps effect. The sample estimator given in (6.3.7) is shown to be biased at high frequencies caused by higher probability for zero valued products of pairwise returns (Section 6.3.3). With the availability of high frequency data, also known as *tick data*, a new line of research has spawned to develop better estimators to measure the realized volatility of an asset or the covariance of two or more assets. There are two major challenges in using high frequency data for covariance estimation. Namely,

1. Market data samples (ticks) of an asset are randomly spaced in time and transactions of different assets occur asynchronously [112].
2. Naturally, there is significant microstructure noise in the market data [92]. In today's markets, order cancellation rates of 80% or more are quite common [111]. Moreover, the negative impact of some HFT practices such as *stuffing* on the markets manifests itself as a very low signal-to-noise ratio in the high frequency data (Section 6.4.4).

The covariance estimators developed for high frequency data can be grouped into two categories as *non-parametric* and *parametric* estimators. Most nonparametric estimation methods for realized variance create a synchronized pair of asset prices by subsampling (using linear interpolation, previous-tick interpolation, or others) and construct an estimator for the synchronized samples [92]. It is known that these estimators have bias, and they need to be calibrated accordingly in order to be useful in practice [113, 114]. Some well-known non-parametric estimators are discussed in detail in [92, 115–118]. In contrast, examples of popular parametric estimators for realized covariance include the Maximum-Likelihood Estimator (MLE) [119] and Quasi MLE (QMLE) estimator [120]. A synchronization scheme reported in [112] generalizes the method introduced in [121] (then available as a technical report). It leverages this synchronization scheme in QMLE to develop an estimator free of tuning parameters and readily implementable. The quest for developing covariance estimators that are robust and resilient to the market microstructure noise continues and still pursued by researchers. Further discussion on the topic may be found in [122].

6.4.3 Low Latency (Ultra-High Frequency) Trading

High frequency trading is a loosely used term. In a certain context, any intraday trading frequency, even trading at 15-min intervals, may be considered as high frequency. However, in today's markets, there are HFT traders getting in and out of trades within a very small fraction of a second [123]. We define the latter as *low latency* or *ultra-high frequency trading* (UHFT) in order to differentiate it from *intraday trading*. Latency in HFT systems is roughly defined as the time it takes to detect the event, to process the event, and to send an order to the exchange in response to the event. Assuming that two HFT competitors experience identical latency to the exchange, the one with the lower latency on its end has the advantage over the other one in such a scenario. In order to reduce the data latency between their systems and the exchange, HFT firms usually pay venues to co-locate their computer servers on premises. It is also known that some use microwave networks rather than optical fiber cables due to its lower latency [124]. To reduce the latency on their end, HFT firms not only deploy the fastest computing and networking equipment available in the market but also invest in custom design systems for fastest possible processing of network packets with market data through programmable *network interface cards* (NIC) leveraging various techniques such as kernel bypass, TCP bypass, TCP offload, remote direct memory access (RDMA), interrupt mitigation, and other available *information technology* (IT) solutions [125–128]. Due to this paradigm shift in electronic trading, more and more high performance digital signal processing (HP-DSP), computer, and high-speed network engineers are employed in the financial industry.

The need for computing devices (and algorithms that run on them) other than *central processing unit* (CPU) and programmable NIC, such as *field-programmable gate array* (FPGA) and *graphics processing unit* (GPU), has been increasing tremendously along with the speed of trading. With the growing need comes the flourishing literature on building parallel versions of known numerical algorithms that are widely used in financial models and algorithmic trading. An application specific hardware designed with an FPGA that works in the network layer is presented in [129]. It is claimed in the paper that four times latency reduction is achieved compared to traditional software-based frameworks. An IP core library (set of functions and circuits that are tested, optimized, and portable) for FPGAs is reported in [130], where an example application using the library that can sustain 10 Gb/s network line rate with a fixed end-to-end latency of 1 microsecond is also included. An event processing engine based on FPGAs with an order of magnitude latency improvement over software based engines is introduced in [131]. GPU implementations to estimate the Hurst exponent and autocorrelation function of a financial time series data is shown in [132]. A discussion on use of CPUs and GPUs for stream aggregation of high frequency data is given in [133]. It is concluded in the article that GPUs have high computation potential but memory transfer is a bottleneck. There are studies on improving memory access and RDMA in GPU [134]. FPGA and GPU implementations of eigenfiltering of the correlation matrix for risk management (Section 5.1) are discussed in [135] and [31], respectively.

6.4.4 Impact of HFT on the Markets

HFT has been the subject of intense debate for the last few years, mostly due to the "Flash Crash" of May 6, 2010, when an extreme volatility observed in the U.S. markets where the major indices (majority of the assets along with them) have fallen almost 5% and bounced back within 30 min. Audit-trail data is studied and concluded that "HFT did not trigger the Flash Crash, but their responses to the unusually large selling pressure on that day exacerbated market volatility" [14]. The reason behind the Flash Crash is further studied and concluded that "order flow toxicity" observed that day that led to the "Flash Crash" can also happen in the future [136].

There'is no strong consensus as of yet on whether HFT has a positive or negative impact on the markets. Most people believe that HFT companies have advantage over average investors since only a small percentage of traders can afford such advanced technology to trade at those very high speeds. Some people think HFT is the number one reason for the increased volatility. It is not uncommon to hear remarks stating that HFT is putting market integrity and stability at risk. On the other hand, advocates of HFT claim that it increases the liquidity in the market, reduces the transaction costs, and contributes to the price discovery process. Hence, HFT makes markets more efficient. However, unlike traditional market makers that are required to provide liquidity, HFT traders are not under such an obligation. Therefore, they can decide to leave the market at any moment, and leaving it alone in the state of self-destructive illiquidity. Moreover, the quotes HFT traders post (limit orders they place) are barely accessible to the majority of the market participants since they cancel a large portion of it within fractions of a second [111]. Some questionable trading practices can only be achieved by HFT traders. They include *quote stuffing* where a trader places and cancels a large number of orders just to create noise and slow the opponents, *layering* in which a trader places one order in a dark pool and a counter order in a regular exchange lower than the best bid/offer in order to attract liquidity for the former, *smoking* where a trader places alluring limit orders and quickly revises them to less generous terms to take advantage of slow traders' order flow [137]. These practices contribute to the negative public image of the HFT. They are considered as manipulative and there have been cases with charges and sanctions imposed on such traders by the regulatory bodies.[2] Another controversial issue is the *flash trading* that allows extra fee paying parties to see the order flow of other market participants before they are posted in an exchange [138].

Majority of the empirical literature supports the idea that HFT improves the market efficiency. A large dataset of 26 HFT firms (participating in 74% of all trades in the U.S. equities market) is analyzed and several findings on the positive impact of HFT firms are reported in [139]. They are summarized as (a) HFT traders contribute significantly to the price discovery (which leads to price stabilization) since they follow price reversal strategies driven by order imbalances, (b) HFT traders do not systematically front run non-HFT firms, (c) HFT traders invest in a less diverse set of strategies than non-HFT firms, (d) HFT traders do not seem to increase volatility and they are possibly reducing it. In [140], a state space model to decompose price movements into permanent and temporary components using an HFT quote and trade dataset is investigated. Those permanent and temporary components are interpreted as information and noise (transitory volatility, pricing error, etc.), respectively. It is concluded in the paper that overall HFT plays a positive role in price efficiency since marketable HFT orders are in the direction of permanent and in the opposite direction of temporary components. Another empirical study also concludes that HFTs decrease spreads and lowers short-term volatility [123]. Similar studies provide supporting evidence that HFT activity is improving the market efficiency [141, 142]. There are also empirical studies advocating that HFT actually has a negative impact on market efficiency [143, 144].

A market with zero bid-ask spread and infinite liquidity is modeled and stated that HFT traders have abnormal profit opportunities in the expense of ordinary traders due to their speed, and they also increase the volatility [145]. Similar theoretical studies are also reported in [146–148]. The readers of more interest may consult [149] and references therein for further discussions.

[2] See Finra news release on September 13, 2010 titled "FINRA Sanctions Trillium Brokerage Services, LLC, Director of Trading, Chief Compliance Officer, and Nine Traders $2.26 Million for Illicit Equities Trading Strategy" and Wall Street Journal article on October 2, 2014 titled "High-Frequency Trader Charged With Market Manipulation."

6.5 SUMMARY

Once the trading strategy generates a signal to open a position and risk management method confirms to place an order, then, it is up to the order execution engine to execute the order such a way that the market impact of the execution is minimized. The most widely used order execution techniques are called the time-weighted average price (TWAP) and volume-weighted average price (VWAP). More sophisticated order execution strategies minimize the market impact for a given execution risk through execution trajectories. Once the trajectory is defined, the next step is to decide whether to place a market order or a limit order. If it is the latter, depending on the state of the limit order book (LOB), the limit price for the order must be defined. Study of LOB to develop more intelligent statistical models and trading strategies than currently available is always a challenging endeavor with potential financial reward. The state of LOB is extremely dynamic since limit and market orders along with order cancellations arrive at very high speeds. The increase in trading frequency makes empirical correlation matrix almost obsolete due to the Epps effect. Improved estimators still using high frequency data to measure the correlations and asset co-movements are available in the literature. High frequency trading (HFT) methods seek small profits per trade and many trades in a day and deliver impressive P&Ls. There are many HFT strategies (mostly proprietary trade secrets) that include market making, order flow detection, news analysis, and arbitrage. Another group of HFT methods includes the low-latency trading where the race is all about reducing the time delay between the detection of an event and placing an order by sophisticated algorithms running on customized operating systems (light OS) with special hardware and high-speed data networks.

CHAPTER 7

Conclusion

In this book, we presented the fundamentals of financial engineering along with their explanations and interpretations from an engineering perspective. Most popular financial instruments such as stocks, options, forward and futures contracts, ETFs, currency pairs, and fixed income securities and their roles are discussed. We covered the basic concepts of quantitative finance including continuous- and discrete-time price formation models with constant and stochastic volatilities as well as jumps, return process of assets in a portfolio and its statistical properties, modern portfolio theory, capital asset pricing model, relative value and factor models, and a widely used factor known as the eigenportfolio, and many others. We highlighted the difference between investing and trading, and terms used in trading such as getting in and out of a position, leverage, going long and short in an asset, and other relevant ones. We delved into the three most commonly employed financial trading strategies. Namely, they are pairs trading, statistical arbitrage, and trend following. We emphasized in a chapter how to estimate portfolio risk, and how to remove the market noise in the empirical correlation matrix of asset returns by using eigenanalysis used in risk management. We also revisited techniques to speed up the risk estimation through approximations. We introduced algorithmic trading, a practice of using algorithms to optimize the execution of orders in order to reduce their market impact. The limit order book of an asset and how market and limit orders along with order cancellations shape it, surveyed through studies that model the book are explained in detail. Insights to explain reasons of Epps effect that highly impact performance of covariance based trading strategies in high frequencies are presented. We finalized our discussion with an extensive survey of high frequency trading (HFT) methods including publicly known strategies that HFT practitioners employ and their impact on the financial markets.

The book attempts to provide a reader friendly introduction of financial engineering and quantitative finance topics. It avoids theoretical rigor and rather clarifies concepts and their reasoning through explanations and examples provided in each chapter. Moreover, the book has a long list of

A Primer for Financial Engineering. http://dx.doi.org/10.1016/B978-0-12-801561-2.00007-1

references to complement its content and purpose. In addition to the references cited in the book, we encourage interested readers for further readings on the subject. For example, we recommend several books including *Risk and Asset Allocation* by Meucci [150], *Financial Modelling with Jump Processes* by Cont and Tankov [20], *An Introduction to High-Frequency Finance* by Dacorogna et al. [113], *Statistical Arbitrage: Algorithmic Trading Insights and Techniques* by Pole [4], and *Financial Signal Processing and Machine Learning* by Akansu et al. [1]. The list of journals published in the field includes *Quantitative Finance* (Routledge), *Journal on Financial Mathematics* (SIAM), *Applied Mathematical Finance* (Routledge), and *Review of Quantitative Finance and Accounting* (Springer). Moreover, there are several financial engineering conferences and workshops being held annually around the globe.

High performance digital signal processing engineers and data scientists well trained and equipped to use, develop, and implement analytical and numerical tools for data intensive applications running on big data IT infrastructure are expected to become a more visible professional group in the frontiers of the financial industry in the coming years. The dramatic penetration of technology and new practices in the sector, in particular, HFT, real-time risk management, and global integration of exchanges and investment activity, have been transforming the financial industry. It is predicted by almost anyone on the Street that this trend will continue faster than ever in the foreseeable future, and financial engineers are the most likely ones to fill the void.

REFERENCES

[1] A.N. Akansu, S.R. Kulkarni, D. Malioutov (Eds.), Financial Signal Processing and Machine Learning, Wiley-IEEE Press, New York, 2016.

[2] Special Issue on Signal Processing for Financial Applications, IEEE Signal Processing Magazine, September 2011, URL http://ieeexplore.ieee.org/xpl/tocresult.jsp?isnumber=5999554&punumber=79.

[3] Special Issue on Signal Processing Methods in Finance and Electronic Trading, IEEE Journal of Selected Topics in Signal Processing, August 2012, URL http://ieeexplore.ieee.org/xpl/tocresult.jsp?isnumber=6239656.

[4] A. Pole, Statistical Arbitrage: Algorithmic Trading Insights and Techniques, John Wiley & Sons, New York, 2008.

[5] T. Berger, Rate-Distortion Theory, John Wiley and Sons, Inc., New York, 2003, ISBN 9780471219286.

[6] F. Black, M.S. Scholes, The pricing of options and corporate liabilities, J. Polit. Econ. 81 (3) (1973) 637-654.

[7] S.L. Heston, A closed-form solution for options with stochastic volatility with applications to bond and currency options, Rev. Financ. Stud., 6 (2) (1993) 327-343, ISSN 08939454, doi:10.2307/2962057.

[8] W.E. Sterk, Comparative performance of the Black-Scholes and Roll-Geske-Whaley option pricing models, J. Financ. Quant. Anal., 18 (3) (1983) 345-354.

[9] M. Avellaneda, S. Zhang, Path-dependence of leveraged ETF returns, SIAM J. Financ. Math. 1 (1) (2010) 586-603.

[10] A. Lipton, Mathematical Methods for Foreign Exchange: A Financial Engineer's Approach, World Scientific, Singapore, 2001.

[11] L. Bachelier, Théorie de la spéculation, Annales scientifiques de l'École Normale Supérieure 3 (1900) 21-86, URL http://www.numdam.org/item?id=ASENS_1900_3_17__21_0.

[12] K. Itô, On a stochastic integral equation, Proc. Jpn. Acad. 22 (2) (1946) 32-35.

[13] A. Papoulis, Probability, Random Variables, and Stochastic Processes, McGraw-Hill, New York, NY, 1991, ISBN 9780073660110.

[14] A.A. Kirilenko, A.S. Kyle, M. Samadi, T. Tuzun, The Flash Crash: the impact of high frequency trading on an electronic market, SSRN eLibrary (2010), URL http://ssrn.com/paper=1686004.

[15] B. Dupire, Pricing with a smile, Risk 7 (1994) 18-20.

[16] E. Derman, I. Kani, Riding on a smile, Risk, 7 (1994) 32-39.

[17] J. Hull, A. White, The pricing of options on assets with stochastic volatilities, J. Finance, 42 (2) (1987) 281-300, ISSN 00221082.

[18] J.C. Cox, J.E. Ingersoll, Jr., S.A. Ross, A theory of the term structure of interest rates, Econometrica, 53 (2) (1985) 385-407, ISSN 00129682.

[19] H.M. Markowitz, Portfolio selection: efficient diversification of investments, Wiley, New York, NY, 1959, x, 344pp.

[20] R. Cont, P. Tankov, Financial Modelling with Jump Processes, CRC Press LLC, Boca Raton, FL, 2003, ISBN 9781584884132.

[21] W.F. Sharpe, Capital asset prices: a theory of market equilibrium under conditions of risk, J. Finance 19 (3) (1964) 425-442.

[22] J.L. Treynor, Market Value, Time, and Risk, Unpublished Manuscript, 1961, pp. 95-209.

[23] J. Lintner, The valuation of risk assets and the selection of risky investments in stock portfolios and capital budgets, Rev. Econ. Stat. (1965) 13-37.

[24] J. Mossin, Equilibrium in a capital asset market, Econometrica: J. Econometric Soc. (1966) 768-783.

[25] J. Ericsson, S. Karlsson, Choosing factors in a multifactor asset pricing model: a Bayesian approach, Tech. Rep., SSE/EFI Working Paper Series in Economics and Finance, 2003.

[26] E.F. Fama, K.R. French, Common risk factors in the returns on stocks and bonds, J. Financ. Econ., 33 (1) (1993) 3-56.

[27] A.N. Akansu, R.A. Haddad, Multiresolution Signal Decomposition: Transforms, Subbands, and Wavelets, Academic Press, Inc., San Diego, CA, 1992, ISBN 012047140X.

[28] I.T. Jolliffe, Principal Component Analysis, Springer-Verlag, New York, NY, 2002.

[29] M. Avellaneda, J.-H. Lee, Statistical arbitrage in the US equities market, Quant. Finance 10 (2010) 761-782.

[30] A.N. Akansu, M.U. Torun, Toeplitz approximation to empirical correlation matrix of asset returns: a signal processing perspective, J. Sel. Top. Signal Process., 6 (4) (2012) 319-326.

[31] M.U. Torun, A.N. Akansu, A novel GPU implementation of eigenanalysis for risk management, in: IEEE 13th International Workshop on Signal Processing Advances in Wireless Communications (SPAWC), 2012, pp. 490-494.

[32] M.U. Torun, A.N. Akansu, An efficient method to derive explicit KLT kernel for first-order autoregressive discrete process, IEEE Trans. Signal Process. 61 (15) (2013) 3944-3953.

[33] B.-L. Zhang, R. Coggins, M.A. Jabri, D. Dersch, B. Flower, Multiresolution forecasting for futures trading using wavelet decompositions, IEEE Trans. Neural Netw. 12 (4) (2001) 765-775.

[34] S.-T. Li, S.-C. Kuo, Knowledge discovery in financial investment for forecasting and trading strategy through wavelet-based SOM networks, Expert Syst. Appl. 34 (2) (2008) 935-951.

[35] S.B. Kotsiantis, I.D. Zaharakis, P.E. Pintelas, Machine learning: a review of classification and combining techniques, Artif. Intell. Rev. 26 (3) (2006) 159-190.

[36] E.M. Azoff, Neural Network Time Series Forecasting of Financial Markets, John Wiley & Sons, Inc., New York, 1994.

[37] I. Kaastra, M. Boyd, Designing a neural network for forecasting financial and economic time series, Neurocomputing, 10 (3) (1996) 215-236.

[38] R.S. Mamon, R.J. Elliott, Hidden Markov Models in Finance, vol. 4, Springer, New York, 2007.

[39] S.-H. Chen, Genetic Algorithms and Genetic Programming in Computational Finance, vol. 1, Springer, New York, 2002.

[40] M. Avellaneda, M. Lipkin, A dynamic model for hard-to-borrow stocks, Risk (2009) 92-97.

[41] M.U. Torun, A.N. Akansu, M. Avellaneda, Portfolio risk in multiple frequencies, IEEE Signal Process. Mag. Spec. Issue Signal Process. Financ. Appl., 28 (5) (2011) 61-71.

[42] G.E. Box, G.M. Jenkins, G.C. Reinsel, Time Series Analysis: Forecasting and Control, John Wiley & Sons, New York, 2013.

[43] R.J. Elliott, J. Van Der Hoek, W.P. Malcolm, Pairs trading, Quant. Finance 5 (3) (2005) 271-276.

[44] B.B. Mandelbrot, J.W. Van Ness, Fractional Brownian motions, fractional noises and applications, SIAM Rev. 10 (4) (1968) 422-437.

[45] N.T. Dung, Fractional geometric mean-reversion processes, J. Math. Anal. Appl. 380 (1) (2011) 396-402.

[46] A.W. Lo, Hedge Funds: An Analytic Perspective, Princeton University Press, Princeton, 2010.

[47] R. Gençay, F. Selçuk, B.J. Whitcher, An Introduction to Wavelets and Other Filtering Methods in Finance and Economics, Academic Press, San Diego, 2001.

[48] J.-H. Wang, J.-Y. Leu, Stock market trend prediction using ARIMA-based neural networks, in: IEEE International Conference on Neural Networks, vol. 4, 1996, pp. 2160-2165.

[49] E.W. Saad, D.V. Prokhorov, D.C. Wunsch, Comparative study of stock trend prediction using time delay, recurrent and probabilistic neural networks, IEEE Trans. Neural Netw., 9 (6) (1998) 1456-1470.

[50] E.G. de Souza e Silva, L.F. Legey, E.A. de Souza e Silva, Forecasting oil price trends using wavelets and hidden Markov models, Energy Econ. 32 (6) (2010) 1507-1519.

[51] A.W. Lo, H. Mamaysky, J. Wang, Foundations of technical analysis: computational algorithms, statistical inference, and empirical implementation, J. Finance, 55 (4) (2000) 1705-1770.

[52] C.-H. Park, S.H. Irwin, What do we know about the profitability of technical analysis?, J. Econ. Surv., 21 (4) (2007) 786-826.

[53] S.K. Mitra, Digital Signal Processing: A Computer-based Approach, 3rd Ed., McGraw-Hill, New York, NY, 2005.

[54] J.F. Ehlers, Cybernetic Analysis for Stocks and Futures: Cutting-Edge DSP Technology to Improve Your Trading, vol. 202, John Wiley & Sons, New York, 2004.

[55] L. Laloux, P. Cizeau, M. Potters, J.-P. Bouchaud, Random matrix theory and financial correlations, Int. J. Theor. Appl. Finance, 3 (2000) 391-397.

[56] V. Plerou, P. Gopikrishnan, B. Rosenow, L.A.N. Amaral, T. Guhr, H.E. Stanley, Random matrix approach to cross correlations in financial data, Phys. Rev. E 65 (2002) 066126-1-066126-18.

[57] J.P. Bouchaud, M. Potters, Financial Applications of Random Matrix Theory: A Short Review, Quantitative Finance Papers, no. 0910.1205, arXiv.org, 2009, accessed on 4/30/2013, URL http://ideas.repec.org/p/arx/papers/0910.1205.html.

[58] A.M. Sengupta, P.P. Mitra, Distributions of singular values for some random matrices, Phys. Rev. E 60 (3) (1999) 3389-3392, doi:10.1103/PhysRevE.60.3389.

[59] T.W. Epps, Comovements in stock prices in the very short run, J. Am. Stat. Assoc. 74 (366) (1979) 291-298, ISSN 01621459.

[60] B. Atal, M.R. Schroeder, Predictive coding of speech signals and subjective error criteria, IEEE Trans. Acoust. Speech Signal Process. ASSP-27 (3) (1979) 247-254.

[61] S. Kay, Modern Spectral Estimation: Theory and Application, Prentice Hall, Upper Saddle River, NJ, 1988.

[62] D. Mueller-Gritschneder, H. Graeb, U. Schlichtmann, A successive approach to compute the bounded Pareto front of practical multiobjective optimization problems, SIAM J. Optim., 20 (2) (2009) 915-934.

[63] K. Deb, Multi-Objective Optimization Using Evolutionary Algorithms, John Wiley & Sons, West Sussex, UK, 2001.

[64] A. Kraus, H.R. Stoll, Price impacts of block trading on the New York Stock Exchange, J. Finance 27 (3) (1972) 569-588.

[65] R.W. Holthausen, R.W. Leftwich, D. Mayers, The effect of large block transactions on security prices: a cross-sectional analysis, J. Financ. Econ. 19 (2) (1987) 237-267.

[66] R.W. Holthausen, R.W. Leftwich, D. Mayers, Large-block transactions, the speed of response, and temporary and permanent stock-price effects, J. Financ. Econ., 26 (1) (1990) 71-95.

[67] R. Almgren, N. Chriss, Optimal execution of portfolio transactions, J. Risk 3 (2001) 5-40.

[68] A.F. Perold, The implementation shortfall: paper versus reality, J. Portf. Manag. 14 (3) (1988) 4-9.

[69] D. Bertsimas, A.W. Lo, Optimal control of execution costs, J. Financ. Mark. 1 (1) (1998) 1-50.

[70] R.C. Grinold, R.N. Kahn, Active Portfolio Management, McGraw-Hill, New York, NY, 2000.

[71] J. Gatheral, No-dynamic-arbitrage and market impact, Quant. Finance, 10 (7) (2010) 749-759.

[72] A. Alfonsi, A. Fruth, A. Schied, Optimal execution strategies in limit order books with general shape functions, Quant. Finance 10 (2) (2010) 143-157.

[73] A. Alfonsi, A. Schied, A. Slynko, Order book resilience, price manipulation, and the positive portfolio problem, SIAM J. Financ. Math. 3 (1) (2012) 511-533.

[74] P.A. Forsyth, J.S. Kennedy, S.T. Tse, H. Windcliff, Optimal trade execution: a mean quadratic variation approach, J. Econ. Dyn. Control, 36 (12) (2012) 1971-1991.

[75] A.A. Obizhaeva, J. Wang, Optimal trading strategy and supply/demand dynamics, J. Financ. Mark., 16 (1) (2013) 1-32.

[76] R. Cont, Statistical modeling of high-frequency financial data, IEEE Signal Process. Mag. 28 (5) (2011) 16-25.

[77] R. Cont, S. Stoikov, R. Talreja, A stochastic model for order book dynamics, Oper. Res., 58 (3) (2010) 549-563.

[78] L.E. Harris, V. Panchapagesan, The information content of the limit order book: evidence from NYSE specialist trading decisions, J. Financ. Mark. 8 (1) (2005) 25-67.

[79] E. Smith, J.D. Farmer, L. Gillemot, S. Krishnamurthy, Statistical theory of the continuous double auction, Quant. Finance, 3 (6) (2003) 481-514.

[80] P. Hewlett, Clustering of order arrivals, price impact and trade path optimisation. in: Workshop on Financial Modeling with Jump processes, Ecole Polytechnique, 2006, pp. 6-8.

[81] R. Cont, A. De Larrard, Price dynamics in a Markovian limit order market, SIAM J. Financ. Math. 4 (1) (2013) 1-25.

[82] A.G. Hawkes, Spectra of some self-exciting and mutually exciting point processes, Biometrika 58 (1) (1971) 83-90.

[83] J.-P. Bouchaud, M. Mézard, M. Potters, et al., Statistical properties of stock order books: empirical results and models, Quant. Finance 2 (4) (2002) 251-256.

[84] M. Potters, J.-P. Bouchaud, More statistical properties of order books and price impact, Physica A: Stat. Mech. Appl. 324 (1) (2003) 133-140.

[85] J. Hasbrouck, Empirical Market Microstructure: The Institutions, Economics, and Econometrics of Securities Trading, Oxford University Press, Oxford, 2006.

[86] A. Alfonsi, A. Schied, Optimal trade execution and absence of price manipulations in limit order book models, SIAM J. Financ. Math. 1 (1) (2010) 490-522.

[87] E. Bayraktar, M. Ludkovski, Liquidation in limit order books with controlled intensity, Math. Finance (2012).

[88] M. Avellaneda, S. Stoikov, High-frequency trading in a limit order book, Quant. Finance 8 (3) (2008) 217-224.

[89] F. Guilbaud, H. Pham, Optimal high-frequency trading with limit and market orders, Quant. Finance 13 (1) (2013) 79-94.

[90] M.U. Torun, A.N. Akansu, On basic price model and volatility in multiple frequencies, in: IEEE Statistical Signal Processing Workshop (SSP), 2011, pp. 45-48.

[91] E. Bacry, S. Delattre, M. Hoffmann, J.F. Muzy, Modeling microstructure noise using Hawkes processes. in: IEEE International Conference on Acoustics Speech and Signal Processing, 2011.

[92] L. Zhang, Estimating covariation: Epps effect, microstructure noise, J. Econ. 160 (1) (2011) 33 - 47, ISSN 0304-4076.

[93] T. Ho, H.R. Stoll, Optimal dealer pricing under transactions and return uncertainty, J. Financ. Econ., 9 (1) (1981) 47-73.

[94] M. O'hara, Market Microstructure Theory, vol. 108, Blackwell, Cambridge, MA, 1995.

[95] H.R. Stoll, Market microstructure, in: Handbook of the Economics of Finance, vol. 1, 2003, pp. 553-604.

[96] B. Biais, L. Glosten, C. Spatt, Market microstructure: a survey of microfoundations, empirical results, and policy implications, J. Financ. Mark. 8 (2) (2005) 217-264.

[97] O. Guéant, C.-A. Lehalle, J. Fernandez-Tapia, Dealing with the inventory risk: a solution to the market making problem, Math. Financ. Econ. 7 (4) (2013) 477-507.

[98] J.Y. Campbell, T. Ramadorai, A. Schwartz, Caught on tape: institutional trading, stock returns, and earnings announcements, J. Financ. Econ. 92 (1) (2009) 66-91.

[99] R. Cont, A. Kukanov, et al., Optimal Order Placement in Limit Order Markets, Available at SSRN 2155218, 2012.

[100] Y. Nevmyvaka, Y. Feng, M. Kearns, Reinforcement learning for optimized trade execution, in: The 23rd International Conference on Machine Learning, 2006, pp. 673-680.

[101] K. Ganchev, Y. Nevmyvaka, M. Kearns, J.W. Vaughan, Censored exploration and the dark pool problem, Commun. ACM 53 (5) (2010) 99-107.

[102] A. Agarwal, P.L. Bartlett, M. Dama, Optimal allocation strategies for the dark pool problem. in: International Conference on Artificial Intelligence and Statistics, 2010, pp. 9-16.

[103] S. Laruelle, C.-A. Lehalle, G. Pages, Optimal split of orders across liquidity pools: a stochastic algorithm approach, SIAM J. Financ. Math. 2 (1) (2011) 1042-1076.

[104] C. Maglaras, C.C. Moallemi, H. Zheng, Optimal order routing in a fragmented market, Preprint, 2012.

[105] D. Easley, M.L. de Prado, M. O'Hara, High-Frequency Trading: New Realities for Traders, Markets and Regulators, Incisive Media, 2013, ISBN 9781782720096, URL https://books.google.com/books?id=pli5oAEACAAJ.

[106] A. Groß-Klußmann, N. Hautsch, When machines read the news: using automated text analytics to quantify high frequency news-implied market reactions, J. Empir. Finance 18 (2) (2011) 321-340.

[107] J. Bollen, H. Mao, X. Zeng, Twitter mood predicts the stock market, J. Comput. Sci., 2 (1) (2011) 1-8.

[108] J. Boudoukh, R. Feldman, S. Kogan, M. Richardson, Which news moves stock prices? A textual analysis, Tech. Rep., National Bureau of Economic Research, 2013.

[109] G. Birz, J.R. Lott Jr., The effect of macroeconomic news on stock returns: new evidence from newspaper coverage, J. Bank. Finance 35 (11) (2011) 2791-2800.

[110] H. Mao, S. Counts, J. Bollen, Predicting financial markets: comparing survey, news, twitter and search engine data, arXiv preprint arXiv:1112.1051, 2011.

[111] M. Chlistalla, B. Speyer, S. Kaiser, T. Mayer, High-frequency trading, in: Deutsche Bank Research, 2011, pp. 1-19.

[112] Y. Aït-Sahalia, J. Fan, D. Xiu, High-frequency covariance estimates with noisy and asynchronous financial data, J. Am. Stat. Assoc., 105 (492) (2010) 1504-1517.

[113] M.M. Dacorogna, R. Gencay, U. Muller, R.B. Olsen, O.V. Pictet, An Introduction to High-Frequency Finance, 2001, Academic Press, New York.

[114] T. Hayashi, N. Yoshida, et al., On covariance estimation of non-synchronously observed diffusion processes, Bernoulli 11 (2) (2005) 359-379.

[115] L. Zhang, P.A. Mykland, Y. Aït-Sahalia, A tale of two time scales, J. Am. Stat. Assoc. 100 (472) (2005).

[116] L. Zhang, et al., Efficient estimation of stochastic volatility using noisy observations: a multi-scale approach, Bernoulli 12 (6) (2006) 1019-1043.

[117] O.E. Barndorff-Nielsen, P.R. Hansen, A. Lunde, N. Shephard, Designing realized kernels to measure the ex post variation of equity prices in the presence of noise, Econometrica 76 (6) (2008) 1481-1536.

[118] J. Jacod, Y. Li, P.A. Mykland, M. Podolskij, M. Vetter, Microstructure noise in the continuous case: the pre-averaging approach, Stoch. Process. Appl. 119 (7) (2009) 2249-2276.

[119] Y. Aït-Sahalia, P.A. Mykland, L. Zhang, How often to sample a continuous-time process in the presence of market microstructure noise, Rev. Financ. Stud. 18 (2) (2005) 351-416.

[120] D. Xiu, Quasi-maximum likelihood estimation of volatility with high frequency data, J. Econ. 159 (1) (2010) 235-250.

[121] O.E. Barndorff-Nielsen, P.R. Hansen, A. Lunde, N. Shephard, Multivariate realised kernels: consistent positive semi-definite estimators of the covariation of equity prices with noise and non-synchronous trading, J. Econ., 162 (2) (2011) 149-169.

[122] P.A. Mykland, L. Zhang, The econometrics of high frequency data, Stat. Methods Stoch. Diff. Equ., 124 (2012) 109.

[123] J. Hasbrouck, G. Saar, Low-latency trading, J. Financ. Markets 16 (4) (2013) 646-679.

[124] E. Budish, P. Cramton, J. Shim, The high-frequency trading arms race: frequent batch auctions as a market design response, 2013.

[125] P. Shivam, P. Wyckoff, D. Panda, EMP: zero-copy OS-bypass NIC-driven gigabit Ethernet message passing, in: Supercomputing, ACM/IEEE 2001 Conference, 2001, pp. 49-49.

[126] K.E. Law, A. Saxena, Scalable design of a policy-based management system and its performance, IEEE Commun. Mag., 41 (6) (2003) 72-79.

[127] D. Goldenberg, M. Kagan, R. Ravid, M.S. Tsirkin, Zero copy sockets direct protocol over infiniband-preliminary implementation and performance analysis. in: 13th Symposium on High Performance Interconnects, 2005, pp. 128-137.

[128] H.-Y. Kim, S. Rixner, TCP offload through connection handoff. in: ACM SIGOPS Operating Systems Review, vol. 40, 2006, pp. 279-290.

[129] C. Leber, B. Geib, H. Litz, High frequency trading acceleration using FPGAs. in: International Conference on Field Programmable Logic and Applications, 2011, pp. 317-322.

[130] J.W. Lockwood, A. Gupte, N. Mehta, M. Blott, T. English, K. Vissers, A low-latency library in FPGA hardware for high-frequency trading. in: IEEE 20th Annual Symposium on High-Performance Interconnects, 2012, pp. 9-16.

[131] M. Sadoghi, M. Labrecque, H. Singh, W. Shum, H.-A. Jacobsen, Efficient event processing through reconfigurable hardware for algorithmic trading, Proc. VLDB Endowment 3 (1-2) (2010) 1525-1528.

[132] T. Preis, GPU-computing in econophysics and statistical physics, Eur. Phys. J. Spec. Top. 194 (1) (2011) 87-119.

[133] S. Schneidert, H. Andrade, B. Gedik, K.-L. Wu, D.S. Nikolopoulos, Evaluation of streaming aggregation on parallel hardware architectures. in: Fourth ACM International Conference on Distributed Event-Based Systems, 2010, pp. 248-257.

[134] J. Duato, A.J. Pena, F. Silla, R. Mayo, E.S. Quintana-Ortí, rCUDA: reducing the number of GPU-based accelerators in high performance clusters. in: IEEE International Conference on High Performance Computing and Simulation, 2010, pp. 224-231.

[135] M.U. Torun, O. Yilmaz, A.N. Akansu, FPGA based eigenfiltering for real-time portfolio risk analysis, in: IEEE International Conference on Acoustics, Speech and Signal Processing, 2013, pp. 8727-8731.

[136] D. Easley, M.M.L. De Prado, M. O'Hara, The microstructure of the Flash Crash: flow toxicity, liquidity crashes and the probability of informed trading, J. Portf. Manag., 37 (2) (2011) 118-128.

[137] B. Biais, P. Woolley, The Flip Side: High Frequency Trading, in: Financial World, 2012.

[138] T. Bhupathi, Technology's latest market manipulator-high frequency trading: the strategies, tools, risks, and responses, NCJL & Tech. 11 (2009) 377.

[139] J. Brogaard, High frequency trading and its impact on market quality, Northwestern University Kellogg School of Management Working Paper, 2010, p. 66.

[140] J. Brogaard, T. Hendershott, R. Riordan, High-frequency trading and price discovery, Rev. Financ. Stud. (2014), doi:10.1093/rfs/hhu032.

[141] T. Hendershott, R. Riordan, Algorithmic trading and information, Manuscript, University of California, Berkeley, 2009.

[142] A.P. Chaboud, B. Chiquoine, E. Hjalmarsson, C. Vega, Rise of the machines: algorithmic trading in the foreign exchange market, J. Finance, 69 (5) (2014) 2045-2084.

[143] F. Zhang, High-frequency trading, stock volatility, and price discovery, Available at SSRN 1691679, 2010.

[144] A. Madhavan, Exchange-traded funds, market structure and the Flash Crash, SSRN Electron. J. (2011) 1-33.

[145] R.A. Jarrow, P. Protter, A dysfunctional role of high frequency trading in electronic markets, Int. J. Theor. Appl. Finance 15 (03) (2012).

[146] J. Cvitanic, A.A. Kirilenko, High frequency traders and asset prices, Available at SSRN 1569075, 2010.

[147] A. Gerig, D. Michayluk, Automated Liquidity Provision and the Demise of Traditional Market Making, Tech. Rep., 2010.

[148] P. Hoffmann, A dynamic limit order market with fast and slow traders, J. Financ. Econ. 113 (1) (2014) 156-169.

[149] B. Biais, P. Woolley, High frequency trading, Manuscript, Toulouse University, IDEI, 2011.

[150] A. Meucci, Risk and Asset Allocation, Springer, New York, 2009.

Printed in the United States
By Bookmasters